Probabilistic Approaches to Recommendations

Synthesis Lectures on Data Mining and Knowledge Discovery

Editor
Jiawei Han, *University of Illinois at Urbana-Champaign*
Lise Getoor, *University of Maryland*
Wei Wang, *University of North Carolina, Chapel Hill*
Johannes Gehrke, *Cornell University*
Robert Grossman, *University of Chicago*

Synthesis Lectures on Data Mining and Knowledge Discovery is edited by Jiawei Han, Lise Getoor, Wei Wang, Johannes Gehrke, and Robert Grossman. The series publishes 50- to 150-page publications on topics pertaining to data mining, web mining, text mining, and knowledge discovery, including tutorials and case studies. The scope will largely follow the purview of premier computer science conferences, such as KDD. Potential topics include, but not limited to, data mining algorithms, innovative data mining applications, data mining systems, mining text, web and semi-structured data, high performance and parallel/distributed data mining, data mining standards, data mining and knowledge discovery framework and process, data mining foundations, mining data streams and sensor data, mining multi-media data, mining social networks and graph data, mining spatial and temporal data, pre-processing and post-processing in data mining, robust and scalable statistical methods, security, privacy, and adversarial data mining, visual data mining, visual analytics, and data visualization.

Graph Mining: Laws, Tools, and Case Studies
D. Chakrabarti and C. Faloutsos
2012

Mining Heterogeneous Information Networks: Principles and Methodologies
Yizhou Sun and Jiawei Han
2012

Privacy in Social Networks
Elena Zheleva, Evimaria Terzi, and Lise Getoor
2012

Community Detection and Mining in Social Media
Lei Tang and Huan Liu
2010

Ensemble Methods in Data Mining: Improving Accuracy Through Combining Predictions
Giovanni Seni and John F. Elder
2010

Modeling and Data Mining in Blogosphere
Nitin Agarwal and Huan Liu
2009

Probabilistic Approaches to Recommendations

Nicola Barbieri, Giuseppe Manco, and Ettore Ritacco

ISBN: 978-3-031-00778-1   paperback
ISBN: 978-3-031-01906-7   ebook

DOI 10.1007/978-3-031-01906-7

A Publication in the Springer series
SYNTHESIS LECTURES ON DATA MINING AND KNOWLEDGE DISCOVERY

Lecture #9

Series Editors: Jiawei Han, *University of Illinois at Urbana-Champaign*
               Lise Getoor, *University of Maryland*
               Wei Wang, *University of North Carolina, Chapel Hill*
               Johannes Gehrke, *Cornell University*
               Robert Grossman, *University of Chicago*

Series ISSN

Print 2151-0067   Electronic 2151-0075

Probabilistic Approaches to Recommendations

Nicola Barbieri
Yahoo Labs, Barcelona, Spain

Giuseppe Manco
ICAR-CNR, Rende, Italy

Ettore Ritacco
ICAR-CNR, Rende, Italy

SYNTHESIS LECTURES ON DATA MINING AND KNOWLEDGE DISCOVERY #9

ABSTRACT

The importance of accurate recommender systems has been widely recognized by academia and industry, and recommendation is rapidly becoming one of the most successful applications of data mining and machine learning. Understanding and predicting the choices and preferences of users is a challenging task: real-world scenarios involve users behaving in complex situations, where prior beliefs, specific tendencies, and reciprocal influences jointly contribute to determining the preferences of users toward huge amounts of information, services, and products. Probabilistic modeling represents a robust formal mathematical framework to model these assumptions and study their effects in the recommendation process.

This book starts with a brief summary of the recommendation problem and its challenges and a review of some widely used techniques Next, we introduce and discuss probabilistic approaches for modeling preference data. We focus our attention on methods based on latent factors, such as mixture models, probabilistic matrix factorization, and topic models, for explicit and implicit preference data. These methods represent a significant advance in the research and technology of recommendation. The resulting models allow us to identify complex patterns in preference data, which can be exploited to predict future purchases effectively.

The extreme sparsity of preference data poses serious challenges to the modeling of user preferences, especially in the cases where few observations are available. Bayesian inference techniques elegantly address the need for regularization, and their integration with latent factor modeling helps to boost the performances of the basic techniques.

We summarize the strengths and weakness of several approaches by considering two different but related evaluation perspectives, namely, rating prediction and recommendation accuracy. Furthermore, we describe how probabilistic methods based on latent factors enable the exploitation of preference patterns in novel applications beyond rating prediction or recommendation accuracy.

We finally discuss the application of probabilistic techniques in two additional scenarios, characterized by the availability of side information besides preference data.

In summary, the book categorizes the myriad probabilistic approaches to recommendations and provides guidelines for their adoption in real-world situations.

KEYWORDS

recommender systems, probability, inference, prediction, learning, latent factor models, maximum likelihood, mixture models, topic modeling, matrix factorization, Bayesian modeling, cold start, social networks, influence, social contagion

This book is dedicated to our families.

To Irene, Nicola, Maddalena and Bella. Thanks for your patience and continued support.

To Caterina and Nino. Thanks for your love and encouragement throughout my life.

To Tiziana, Teodora and Francesco for their love.

Contents

Preface

Recommendation is a special form of information filtering that extends the traditional concept of search by modeling and understanding the personal preferences of users. The importance of accurate recommender systems has been widely recognized by both academic and industrial efforts in the last two decades and recommender systems are rapidly becoming one of the most successful applications of data-mining and machine-learning techniques.

Within this context, a rich body of research has been devoted to the study of probabilistic methods for the analysis of the preferences and behavior of users. Real-world scenarios involve users in complex situations, where prior beliefs, specific tendencies, and reciprocal influences jointly contribute to determining the preferences of users toward huge amounts of information, services, and products. Can we build simple yet sophisticated models to explain the way these preferences occur? How can we identify and model the latent causes that guide and influence the adoptions and preferences of users in complex systems? Probabilistic modeling represents a robust, formal mathematical framework upon which to model assumptions and study their effects in real-world scenarios.

In this book, we study and discuss the application of probabilistic modeling to preference data provided by users. We draw from recent research in the field and attempt to describe and abstract the process adopted by the research community in the last five years. The increasing popularity of recommender systems (due also to the advent of the Netflix prize) led to an explosion of research on generative approaches for modeling preference data. Within this scenario, an awareness has grown that probabilistic models offer more than traditional methods in terms of both accuracy and exploitation. Hence the need for a systematic study aimed at categorizing the myriad approaches and providing guidelines for their adoption in real-world situations.

We start by formally introducing the recommendation problem and summarizing well-established techniques for the generation of personalized recommendations. We present an overview of evaluation methods and open challenges in the recommendation process, and then focus on collaborative filtering approaches to recommendation.

The core of the book is the description of probabilistic approaches to recommendations. We concentrate on probabilistic models based on latent factors, which have proved to be extremely powerful and effective in modeling and identifying patterns within high-dimensional (and extremely sparse) preference data. The probabilistic framework provides a powerful tool for the analysis and characterization of complex relationships among users and items. Probabilistic models can enhance and strengthen traditional techniques as they offer some significant advantages. Notably, they can: be tuned to optimize any of several loss functions; optimize the likelihood of enabling the modeling of a distribution over rating values, which can be used to determine the confidence of the model in providing a recommendation; finally, they offer the possibility of including prior knowledge in the generative process, thus allowing a more effective modeling of the underlying data distribution.

Rooted in these backgrounds, this book focuses on the problem of effectively adopting probabilistic models for modeling preference data. Our contributions can be summarized in two ways. First, we study the effectiveness of probabilistic techniques in generating accurate and personalized recommendations. Notably, the interplay of several different factors, such as the estimate of the likelihood that the user will select an item, or the predicted preference value, provides a fruitful degree of flexibility. Secondly, we show how probabilistic methods based on latent factor modeling can effectively capture interesting local and global patterns embedded in the high-dimensional preference data, thereby enabling their exploitation in novel applications beyond rating prediction.

The effectiveness of probabilistic methods for recommendation has also been shown in other scenarios. The first concerns contextual information. Typically, preference data assumes a "bag-of-words" representation, i.e. the order of the items a user chooses can be neglected. This assumption allows us to concentrate on recurrent co-occurrence patterns and to disregard sequential information that may characterize the choice of the user. However, there are several real-world applications where contextual and side information play a crucial role. For example, content features can help characterize the preferential likelihood of an item in extremely dynamic content production systems. Also, there are some scenarios, such as web navigation logs or customer purchase history, where data can be "naturally" interpreted as sequences. Ignoring the intrinsic sequentiality of the data may result in poor modeling. By contrast, the focus should be on the context within which a current user acts and expresses preferences, i.e., the environment, characterized by side information, where the observations hold. The context enables a more refined modeling and, hence, more accurate recommendations.

Finally, the advent of social networking introduces new perspectives in recommendation. Understanding how the adoptions of new practices, ideas, beliefs, technologies and products can spread trough a population, driven by social influence, is a central issue for the social science and recommendation communities. Taking into account the modular structure of the underlying social network provides further important insights into the phenomena known as social contagion or information cascades. In particular, individuals tend to adopt the behavior of their social peers so that information propagation triggers "viral" phenomena, which propagate within connected clusters of people. Therefore, the study of social contagion is intrinsically connected to the recommendation problem.

In this book, the interested reader will find the mathematical tools that enable the understanding of the key aspects that characterize the main approaches for recommendation. This will allow both the understanding of the opportunities of the current literature and the development of ad-hoc methods for the solution of specific problems within the recommendation scenario.

The book is not self-comprehensive and a basic understanding probability and statistics, as well as machine learning and data mining methods, is needed.

The dependencies between chapters are shown in the following figure, where an arrow from one chapter to another indicates that the latter requires some knowledge of the first.

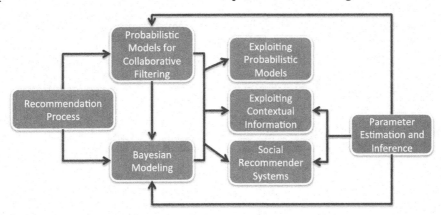

Within this structure, the appendix contains a primer on the basic inference and estimation tools used throughout the chapters and it should be considered as a reference tool.

Nicola Barbieri, Giuseppe Manco, and Ettore Ritacco
April 2014

CHAPTER 1

The Recommendation Process

1.1 INTRODUCTION

With the increasing volume of information, products, services, and resources (or, more generally, items) available on the Web, the role of *recommender systems (RS)* [162] and the importance of highly accurate recommendation techniques have become major concerns both in e-commerce and academic research. The goal of a RS is to provide users with recommendations that are non-trivial, useful to directly experience potentially interesting items, and that the user may not find on her own. Their exploitation in e-commerce enables a better interaction between the users and the system: RSs are widely adopted in different contexts, from music (Last.fm[1]) to books (Amazon[2]), movies (Netflix[3]), and news (Google News[4][49]), and they are quickly changing and reinventing the world of e-commerce [176].

A recommender can be considered as a "push" system that provides users with a personalized exploration of a wide catalog of possible choices. Search-based systems require a user to explicitly express a query specifying what she is looking for. By contrast, in recommendation, the query is implicit and corresponds to all past interactions of the user with the system. By collecting and analyzing past users' preferences in the form of explicit ratings or like/dislike products, the RSs provide the user with smart and personalized suggestions, typically in the form of "customers who bought this item also bought" or "customers who bought related items also bought." The goal of the service provider is not only to transform a regular user into a buyer, but also to make her browsing experience more comfortable, thus building a strong loyalty bond. This idea is better explained by Jeff Bezos, CEO of Amazon.com:

> *"If I have 3 million customers on the Web, I should have 3 million stores on the Web."*

The strategic importance of accurate recommendation techniques has motivated both academic and industrial research for over 15 years; this is witnessed by huge investments in the development of personalized and highly accurate recommendation approaches. In October 2006, Netflix, the leader in the American movie-rental market, promoted a competition, the *Netflix Prize*[5][27], with the goal of producing a 10% improvement on the prediction accuracy achieved by their internal recommender system, *Cinematch*. The competition lasted three years and involved several

---

[1] last.fm
[2] amazon.com
[3] netflix.com
[4] news.google.com
[5] netflixprize.com

research groups from all over the world, improving and inspiring a fruitful research. During this period, a huge number of recommendation approaches were proposed, creating a substantial advance in the literature, while simultaneously encouraging the emergence of new business models based on RSs.

This chapter is aimed at providing an overview of the recommendation scenario from a technical point of view. First, we provide a formal framework that specifies some notations used throughout the manuscript. The focus is the formal definition of a recommender system and the evaluation of its capability to provide adequate recommendations. Then we will introduce and discuss some well-known recommendation approaches, with an emphasis on collaborative filtering.

1.2   FORMAL FRAMEWORK

Recommendation is a form of information filtering that analyzes users' past preferences on a catalog of items to generate a personalized list of suggested items. In the following, we introduce some basic notations to model users, items, and their associated preferences.

Let $\mathcal{U} = \{u_1, \ldots, u_M\}$ be a set of M users and $\mathcal{I} = \{i_1, \ldots, i_N\}$ a set of N items. For notational convenience, in the following, we reserve letters u, v to denote users from $\mathcal{U}$ and letters i, j to indicate items from $\mathcal{I}$. Users and items can be characterized by means of specific properties, specified as atomic predicates in first-order logic. Predicates can be related with either discrete properties (such as Sex(u,male), Location(u,NYC), or Genre(i,comedy), Rating(i,PG) regarding the user u and the item i), or numeric (such as Age(u,23) or Length(i, 175)). Users' preferences can be represented as a $M \times N$ matrix $\mathbf{R}$, whose generic entry r_i^u denotes the preference value (i.e., the degree of appreciation) assigned by user u to item i. User preference data can be classified as *implicit* or *explicit*. Implicit data correspond to mere observations of co-occurrences of users and items, which can be recorded by analyzing clicks, users' web sessions, likes, or check-in. Hence, the generic entry r_i^u of the user-item rating matrix $\mathbf{R}$ is a binary value: $r_i^u = 0$ means that u has not yet experienced i, whereas by $r_i^u = 1$ we denote that user u has been observed to experience item i. By contrast, explicit data record the actual ratings explicitly expressed by individual users on the experienced items. Ratings are typically collected by questionnaires in which a user is asked to proactively evaluate the purchased/selected product on a given rating scale. Such feedback provides a user with a way to explicitly express their preferences. Implicit feedback is abundant but often unreliable, as the true users' evaluations are still hidden. Explicit feedback is generally more accurate and can be either positive or negative, while implicit feedback is always positive.

Generally, explicit ratings can be encoded as scores in a (totally-ordered) numeric domain $\mathcal{V}$, represented as a fixed rating scale that often includes a small number of interestingness levels. In such cases, for each pair (u, i), rating values r_i^u fall within a limited range $\mathcal{V} = \{0, \ldots, V\}$, where 0 represents an unknown rating and V is the maximum degree of preference. Notation $\bar{r}_{\mathbf{R}}$ denotes the average rating among all those ratings $r_i^u > 0$. The number of users M, as well as the number of items N, are very large (typically with $M >> N$) and, in real-world applications, the rating matrix $\mathbf{R}$ is characterized by an exceptional sparseness, as individual users tend to rate a limited number of items. In the following, we denote by $\langle u, i \rangle$ the enumeration of all those dyads in $\mathbf{R}$ such that $r_i^u > 0$; analogously $\langle u, i, r \rangle$ represents an enumeration of all the explicit ratings. Again, specific properties in the form of first-order predicates can be associated with either $\langle u, i \rangle$ or $\langle u, i, r \rangle$. Example properties can specify, e.g., the time period Period($\langle u, i \rangle$, evening), or even the item experienced before Preceding($\langle u, i \rangle$, j). We denote the whole set of predicates relative to $\mathcal{U}, \mathcal{I}$ or $\mathcal{U} \times \mathcal{I} \times \mathcal{V}$ as $\mathcal{F}$.

The set of items rated by user u is denoted by $\mathcal{I}_{\mathbf{R}}(u) = \{i \in \mathcal{I} | \langle u, i \rangle \in \mathbf{R}\}$. Dually, $\mathcal{U}_{\mathbf{R}}(i) = \{u \in \mathcal{U} | \langle u, i \rangle \in \mathbf{R}\}$ is the set of all those users, who rated item i. With an abuse of notation, we use $\mathcal{U}(i)$ and $\mathcal{I}(u)$ when the rating matrix $\mathbf{R}$ is known from the context. Any user u with a rating history, i.e., such that $\mathcal{I}_{\mathbf{R}}(u) \neq \emptyset$, is an *active user*. Both $\mathcal{I}_{\mathbf{R}}(u)$ and $\mathcal{U}_{\mathbf{R}}(i)$ can be empty. This setting is known as *cold start* and it generally occurs whenever a new user or item is added to the underlying information system. Cold start is generally problematic in recommender systems, since these cannot provide suggestions for users or items in the absence of sufficient information.

The illustration in Fig 1.1 sketches a recommendation scenario with 10 users, 10 items and explicit preferences. The set of users who rated item i_2 is $\mathcal{U}(i_2) = \{u_1, u_4, u_8, u_{10}\}$. Also, the set of items rated by user u_2 is $\mathcal{I}(u_2) = \{i_3, i_5, i_7, i_9\}$. The rating value of user u_2 over item i_4, as well as the ratings from u_4 over i_1 and i_3, are unknown.

	i_1	i_2	i_3	i_4	i_5	i_6	i_7	i_8	i_9	i_{10}
u_1	2	5			3	1		2		
u_2			2		4		5		1	
u_3	1		4			5		1		
u_4		4			4		2		2	
u_5	1			2	3	3		5		
u_6			3	5	1	3			4	5
u_7	4			3	5			1	1	
u_8	3	4					1		3	
u_9	1				3	2			4	4
u_{10}		5				5			5	4

Figure 1.1: Example of users' preference matrix.

Given an active user u, the goal of a RS is to provide u with a recommendation list $\mathcal{L}_u \subseteq \mathcal{I}$, including unexperienced items (i.e., $\mathcal{L}_u \cap \mathcal{I}_{\mathbf{R}}(u) = \emptyset$) that are expected to be of her interest. This clearly involves predicting the interest of u toward unrated items: exploiting (implicit and/or explicit) information about users' past actions, the RS provides a scoring function $p_i^u : \mathcal{U} \times \mathcal{I} \to \mathbb{R}$, which accurately estimates future preferences, and hence can be used to predict which are the most likely products to be purchased in the future. A general framework for the generation of $\mathcal{L}_u$, is encoded by algorithm 1. Here, a subset C of candidate items is chosen according to domain-specific criteria. For example, C may correspond to some items to promote, or even to the whole domain $\mathcal{I}$. Then, each item i in C is scored by p_i^u, and the top-L items are selected for the recommendation list.

Algorithm 1 Generation of recommendation list.

Require: A number of candidate items, D (positive integer such that $D \leq N$)
Require: The size of the recommendation list, L (positive integer such that $L \leq D$)
  1: Choose $\mathcal{C} \subseteq \mathcal{I}$, according to a business-specific criterion, such that $|\mathcal{C}| = D$ and $\mathcal{C} \cap \mathcal{I}_{\mathbf{R}}(u) = \emptyset$.
  2: Associate each item $i \in \mathcal{C}$ with a score p_i^u, which represents the appreciation of u for i
  3: Let $\mathcal{L}_u$ be the selection of the top L items in $\mathcal{C}$ with the highest values p_i^u
  4: Return $\mathcal{L}_u$

1.2.1   EVALUATION

Different measures have been proposed in order to evaluate the accuracy of a RS (for a detailed survey see [79, 103]). As mentioned before, recommendations usually come in the form of a list of the L items that the user might like the most. Intuitively, an accuracy metric should measure how close the predicted list is to the actual preference list of a user or how close a predicted rating is to its actual value.

The adoption of a recommender system requires a prior evaluation of its ability to provide a positive impact on both a user's experience and a provider's revenues. Notably, the evaluation is accomplished by applying a protocol that delivers some objective measures stating the quality of the RS. These measures allow for a comparison among the several algorithmic and modeling alternatives. Two different strategies can be pursued, namely *offline* and *online* evaluation. Within online evaluation, a small percentage of users are asked to test the recommendation engine and to provide feedback about the main features. Online testing requires a careful design of the testing protocol, in order for the evaluation to be fair and the results to be statistically sound. This includes the selection of a representative subset of users and the features to be evaluated. Additionally, a direct involvement of actual users is not advisable, as it could negatively affect the experience of the test users: negative judgements can bias a user toward an under-evaluation of the capabilities of the recommendation engine.

By contrast, offline evaluation simulates the online process by employing simple statistical indices (either on synthesized or historical data) to measure the performance of the system. Indices make comparisons among alternative algorithms and schemes easier, and measure the effectiveness of a system at design time. Offline evaluation protocols and metrics usually rely on an evaluation framework where the rating matrix $\mathbf{R}$ is split into two matrices $\mathbf{T}$ and $\mathbf{S}$, such that the former is used to train a recommendation algorithm, while the latter is used for validation purposes, to measure the predictive abilities of the system. The latter can be measured according to different perspectives, discussed in the following.

Prediction accuracy. Historically, most CF techniques focused on the development of accurate techniques for rating prediction. The recommendation problem has been interpreted as a missing value prediction problem [174], in which, given an active user, the system is asked to predict her preference for a set of items. In this respect, the score p_i^u can be devised as a function of the prediction provided by the RS. Predictive accuracy metrics measure how close the predicted ratings are to the actual ratings. Let $\hat{r}_i^u$ denote the predicted rating on the dyad $\langle u, i \rangle$. Prediction accuracy is measured by means of statistical metrics that compare such predicted values with actual ratings. Widely adopted measures are:

- The **Mean Absolute Error (MAE)**, which measures the average absolute deviation between a predicted and a real rating. It is an intuitive metric, easy to compute and widely used in experimental studies:

$$MAE = \frac{1}{|\mathbf{S}|} \sum_{\langle u,i \rangle \in \mathbf{S}} |r_i^u - \hat{r}_i^u|. \tag{1.1}$$

- The **Mean Squared Error (MSE)** and **Root Mean Squared Error (RMSE)**. These measure the deviation of observed ratings from predicted values and, compared to MAE, emphasize large errors. They are defined as

$$MSE = \frac{1}{|\mathbf{S}|} \sum_{\langle u,i \rangle \in \mathbf{S}} \left(r_i^u - \hat{r}_i^u\right)^2, \tag{1.2}$$

and

$$RMSE = \sqrt{MSE}. \tag{1.3}$$

- The **Mean Prediction Error (MPE)**, expresses the percentage of predictions which differ from the actual rating values,

$$MPE = \frac{1}{|\mathbf{S}|} \sum_{\langle u,i \rangle \in \mathbf{S}} \mathbb{1}[\![r_i^u \neq \hat{r}_i^u]\!], \tag{1.4}$$

where $\mathbb{1}[\![bexpr]\!]$ is the indicator function which returns 1 if the boolean expression *bexpr* holds and 0 otherwise. Notably, MPE resembles the classical accuracy measured on classification algorithms in machine learning and data mining. As such, it can be further refined into a confusion matrix denoting the misclassifications relative to each specific rating. Other more refined metrics relying on *ROC analysis* [56] can be devised, based on such a confusion matrix.

Recommendation Accuracy. The focus here is on the recommendation list $\mathcal{L}_u$, and we are interested in measuring the accuracy in building a list that matches the user's preferences. If we consider the problem from an information retrieval (IR) perspective [14], we can assume an explicit knowledge of u's actual preferences through a specific list $\mathcal{T}_u$. Classical IR measures can then be adapted to measure the correspondence between the proposed and the actual list.

Items in $\mathcal{T}_u$ represent the portion of the catalog that are likely to meet u's preferences. With implicit preferences, $\mathcal{T}_u$ can be any subset in $\mathcal{I}_\mathbf{S}$: the assumption here is that the user only selects relevant items. With explicit feedback, the relevance of an item can be further refined by focusing on those items in $\mathcal{I}_\mathbf{S}$ for which the evaluation is representative of high interest. For example, we can fix a threshold t (e.g., the average evaluation $\bar{r}_\mathbf{T}$), and then select $\mathcal{T}_u$ as a random subset of $\{i \in \mathcal{I}_\mathbf{S}(u) | r_i^u > t\}$.

Given $\mathcal{T}_u$ for each user u, the evaluation of a recommendation list built according to Algorithm 1 can be measured through the standard precision and recall measures. In particular, the quality of a system which generates recommendation lists of size L can be measured as

$$
\begin{aligned}
Recall(L) &= \frac{1}{M} \sum_{u \in \mathcal{U}} \frac{|\mathcal{L}_u \cap \mathcal{T}_u|}{|\mathcal{T}_u|}. \\
Precision(L) &= \frac{1}{M} \sum_{u \in \mathcal{U}} \frac{|\mathcal{L}_u \cap \mathcal{T}_u|}{|\mathcal{L}_u|}.
\end{aligned} \tag{1.5}
$$

Precision and recall are conflicting measures. For instance, by increasing the size of the recommendation list, recall is expected to increase but precision decreases. Typically, these measures are combined into the *F-score*, a single measure representing their harmonic mean and balancing both contributions:

$$
F = 2 \cdot \frac{Precision \cdot Recall}{Precision + Recall}. \tag{1.6}
$$

Recall and precision can also be exploited in a more refined evaluation protocol that relies on the notion of *hit rate*. We can devise a random process where the recommendation list is built by scoring a set made of both random elements and elements within $\mathcal{T}_u$. Then, hits represent the elements in $\mathcal{L}_u$ that also belong to $\mathcal{T}_u$. This testing protocol, called *user satisfaction* on the recommendation list, was proposed by Cremonesi et al. in [47] and is illustrated in Algorithm 2. Recall values relative to a user u can be adapted accordingly:

$$
US\text{-}Recall(u, L) = \frac{\#hits}{|\mathcal{T}_u|}, \tag{1.7}
$$

and global values can be obtained by averaging overall users. As for precision, there is a difficulty in considering *false positives*, which negatively influence the accuracy of the system. False positives represent the amount of spurious items, which are not relevant for the user and by contrast are scored high in the list. Clearly, the size L can affect the amount of such items, and hence, a suitable approximation for precision can be obtained by simply reweighting the recall to represent the percentage of relevant items relative to the recommendation list,

$$
US\text{-}Precision(u, L) = \frac{\#hits}{L \cdot |\mathcal{T}_u|}. \tag{1.8}
$$

More refined measures can be introduced that consider both the amount of spurious items and the position of the relevant item within the list.

Algorithm 2 User satisfaction of user u for item i.

Require: An item $i \in \mathcal{T}_u$
Require: A number of candidate items, D (positive integer such that $D \leq |\mathcal{I}|$)
Require: The size of the recommendation list, L (positive integer such that $L \leq D$)
 1: Let $\mathcal{C}$ be a random subset of $\mathcal{I}$, with size D, whose elements have not been rated by u
 2: Add i in $\mathcal{C}$
 3: Assign to each item $i \in \mathcal{C}$ a score p_i^u, which represents the appreciation of u for i
 4: Let $\mathcal{L}_u$ be the selection of the L items in $\mathcal{C}$ with the highest values p_i^u
 5: **if** $i \in \mathcal{L}_u$ **then**
 6:     return a *hit*
 7: **else**
 8:     return a *miss*
 9: **end if**

Rank Accuracy. The generation of a recommendation list is based on the generation of a scoring function. In our basic framework, we assumed that users provide explicit feedback in terms of a scoring value in a given range. However, an explicit feedback can also consist of an explicit ranking of items for each user. That is, users can provide information on which items are better than others, according to their preferences.

True ranking is unrealistic with huge item catalogs. Nevertheless, it allows us to define ranking accuracy metrics that measure the adequacy of the RS in generating a personalized ordering of items that matches with the true user's ordering.

Typically, the predicted and observed orderings are compared by means of *Kendall's* ($\mathcal{K}$) or *Spearman's* (ρ) coefficients. Let τ_u be the full descendingly ordered ranking list for the user u and let $\hat{\tau}_u$ denote the predicted ranking list, both defined over the domain $\mathcal{I}_{\mathbf{S}}(u)$. Then:

$$\mathcal{K}(\tau_u, \hat{\tau}_u) = \frac{2 \left(\sum_{i,j \in \mathcal{I}} \mathbb{1}[\![\tau_u(i) \prec \tau_u(j) \; \wedge \; \hat{\tau}_u(j) \prec \hat{\tau}_u(i)]\!] \right)}{N(N-1)} \tag{1.9}$$

$$\rho(\tau_u, \hat{\tau}_u) = \frac{\sum_{i \in \mathcal{I}} (\tau_u(i) - \bar{\tau}_u)(\hat{\tau}_u(i) - \bar{\hat{\tau}}_u)}{\sqrt{\sum_{i \in \mathcal{I}} (\tau_u(i) - \bar{\tau}_u)^2 \sum_{i \in \mathcal{I}} (\hat{\tau}_u(i) - \bar{\hat{\tau}}_u)^2}} \tag{1.10}$$

where $\tau(i)$ is the rank associated with the item i in the list τ, $\bar{\tau}$ is the average rank in the list τ, and $i \prec j$ is the comparison operator to denote if i is ranked ahead of j. The index $\mathcal{K}$, within the range $[-1, 1]$, measures the agreement between two rankings, which in this case are the real ranked list and the predicted one. If the predicted ranked list matches perfectly with the observed one, then $\mathcal{K} = 1$. By contrast, ρ ranges within $[-1, +1]$ and measures the correlation between two ranked variables. Once again, the perfect matching yields $\rho = 1$.

Other Evaluation Measures.   The aforesaid accuracy metrics are important for offline evaluation and provide viable mathematical frameworks to compare different recommendation algorithms and techniques. However, they are not focused on evaluating the output of a recommender, as users really perceive it from a psychological standpoint. Several key dimensions, beyond accuracy, are gaining increasing importance in the evaluation of the quality of recommendations:

Novelty. This is closely related to the utility of a RS, which should recommend items the user is not already aware of. Recommending a new "Star Trek" movie to a Mr. Spock's fan is not a novel suggestion, since it is likely that the user is already aware of the movie.

Serendipity. This measures the extent to which the RS is able to suggest items that are both of interest to the user and *unexpected, surprising*. Intuitively, assuming that a list of "obvious" items to be recommended is available, serendipity can be measured as the ability of the RS to avoid the recommendations of such items.

Diversity. This measures the capability of suggesting "different" items. Most often, a real RS provides focused suggestions, which are similar to each other, e.g., books by the same author. Diversity should allow recommenders to include items that span, as much as possible, over all of a user's interests.

Coverage. This focuses on the domain of items which the RS is effectively able to recommend to users, and measures whether recommender systems are biased toward some specific subsets.

Limitations of Off-Line Evaluations.   Evaluating recommendations is still among the biggest unsolved problems in RSs. Offline evaluation does not necessarily align with actual results: offline metrics like prediction/recall, RMSE, and $\mathcal{K}$ are based on snapshots of the past activities of users, and they cannot measure the real impact of recommendations on users' purchasing activities. The fact that a recommender matches past preferences does not necessarily mean that it is capable of influencing a user's behavior in the future. A strong issue is how to consider the unrated items. Do users dislike them? Haven't users experienced them yet? Do users know about their existence?

The limitations of off-line evaluation motivates the study of more sophisticated evaluation protocols and accurately controlled experiments. In general, a more effective evaluation should focus on how the user perceives recommendations and on measuring their helpfulness. Current research is investigating several aspects correlated with evaluation, such as the analysis of the *return of investment (ROI)* in improving prediction and recommendation metrics and the proposal of more convincing offline evaluation metrics.

1.2.2   MAIN CHALLENGES

There are a number of issues that the design of a recommendation algorithm is typically required to deal with. These are generally relative to the quality and the size of the data, and the security and privacy issues that a recommendation engine may breach. A brief review of these challenges is summarized next.

Sparsity and Cold Start.   Recommendation systems in e-commerce have to deal with a huge number of items and users. Usually, even the most active users purchase only a limited fraction of the available items. The sparsity problem poses a real challenge to recommendation techniques. For example, the preference matrix might not provide enough information for particular users/items. This problem is known as *reduced coverage* [175], and can seriously affect the accuracy of recommendations. A wide range of methods have been proposed to deal with the sparsity problem. Many of them use dimensionality reduction techniques to produce a low-dimensional representation of the original customer-product space [175].

A related shortcoming is the inclusion of new items or new users. This problem is known as cold-start [177] and has been formalized before. The recommendation engine cannot provide suggestions to unknown users, or regarding unknown items. The cold-start new-user scenario can be faced by soliciting information about her tastes via *preference elicitation*. For example, a new user can be asked to rate a list of items in the registration phase. Selecting the most informative items to propose is a challenge in this context [158], as this selection is crucial in order to obtain an initial estimate of a user's preferences. Other methods focus on the definition of transitive properties between users via trust inferences [146], or by exploiting graph-theory [5].

More generally, cold start can be faced by means of hybrid approaches that consider both the rating matrix and additional information about items and/or users. Demographic information about the user [151] (such as age, sex, marital status) and item features (such as genre, rating, year of release) [154], can be used to infer relationships between the new user (or new item) and already registered users (or items in the catalog) for which the recommendation system already has enough information.

Scalability.   Large user and item bases require robust computational resources. This is especially true in recommender systems associated with e-commerce or web services, like Amazon.com or Google News, which also require real-time response rates. The need to generate high-quality suggestions in a few milliseconds calls for the adoption of scalable methods and sophisticated data structures for data analysis, access, and indexing. Since the number of users is, typically, much larger than the number of items (even different orders of magnitude), scalable approaches focus on the analysis of the item-to-item relationships [174] or tend to model single users in communities [86]. Dimensionality reduction techniques [175], even in incremental versions [173], are used as well, as they allow us to discriminate between a learning phase, to be processed offline, and a recommendation phase, which can be performed online.

Item Churn.   In dynamic systems, such as news recommenders, items are inserted and deleted quickly. The obsolescence of the model in such situations is a big issue, and it calls for the need to develop incremental model-maintenance techniques.

Privacy.   Recommender systems can pose serious threats to individual privacy. Personal data, collected using the interactions of customers with the systems, is commonly sold when companies suffer bankruptcy [38] and, although recommendations are provided in an anonymous form, aggregations of user preferences can be exploited to identify information about a particular user [157]. Sensitive information, like political and sexual orientations, were breached from the dataset made public for the Netflix prize [141].

Privacy-preserving techniques for recommendation include randomized perturbation methods [153] and user-community aggregations of preferences that do not expose individual data [38, 39]. Experimental studies have shown that accurate recommendations are still possible while preserving user privacy.

Security.   Collaborative recommender systems can be exposed to malicious users willing to influence suggestions about items. An attacker can modify the original behavior of the systems by using a set of *attacker profiles* [198], corresponding to fictitious user identities. Three different attack intents have been identified [109]: the goal of an attacker is "pushing" a particular item by raising its predicted ratings; or "nuking" competitor products by lowering predicted ratings; and, in the worst case, damaging the whole recommender system. The works [109, 145, 171] study the robustness of collaborative recommendation, as well as countermeasures for making attacks ineffective. Recent studies [136, 198] have formalized the push/nuke attacks and, consequently, they propose classification techniques for detecting the attacking profile. The general form of an attack is composed of a *target item* (i.e., the item to push or nuke), a set of ratings over *selected items* (chosen for their similarity respect to the target item), and a set of ratings on *filler items*. Each attack can be characterized by considering how the attackers identified the selected items, what proportion of the remaining items they choose as filler items, and how they assign specific ratings to each set of items.

1.3   RECOMMENDATION AS INFORMATION FILTERING

The idea behind information filtering is that human interests and preferences are essentially correlated, thus, it is likely that a target user tends to prefer what other like-minded users have appreciated in the past. Hence, the basic approach to information filtering consists of collecting data about user interests and preferences to build reliable user profiles. These allow us to predict the interests of a target user based on the known preferences of similar users and, eventually, the provision of suitable recommendations of potentially relevant items. Recommendations generated by means of *filtering* techniques can be divided into three main categories.

- **Demographic filtering** approaches assume the possibility of partitioning the overall set of users into a number of classes with specific demographic features. Customization is achieved by exploiting manually-built, static rules that offer recommendations to the target user on the basis of the demographic features of the class she belongs to.

- **Content-based filtering** techniques learn a model of users' interests and preferences by assuming that item and user features are collected and stored within databases' information systems. The aim is to provide the target user with recommendations of unexperienced content items that are thematically similar to those she liked in the past.

- **Collaborative filtering** works by matching the target user with other users with similar preferences and exploits the latter to generate recommendations.

We briefly describe the first two information filtering techniques. Collaborative filtering approaches are the focus of the entire monograph and will be detailed in the next section.

1.3.1 DEMOGRAPHIC FILTERING

Demographic filtering partitions users in $\mathcal{U}$ into a certain number of disjoint classes $\mathcal{U}^{(1)}, \ldots, \mathcal{U}^{(C)}$ and then establishes relationships between items in $\mathcal{I}$ and the corresponding class(es) of interested users. Personal demographic data is basically exploited to group users into classes according to some partitioning criterion, such as the homogeneity of their purchasing history or lifestyle characteristics [108]. A set $\mathcal{P} = \{p^{(1)}, \ldots, p^{(C)}\}$ of handcrafted rules [121] is then exploited to specify the set of items recommendable to the users within the individual classes. A generic rule $p^{(c)}$ associates $\mathcal{U}^{(c)}$ with the corresponding subset $\mathcal{I}^{(c)}$, so that, for a given user u in $\mathcal{U}^{(c)}$, the (binary) prediction criterion

$$\hat{r}_i^u = \begin{cases} 1 & \text{if } i \in p^{(c)}(u) \\ 0 & \text{otherwise} \end{cases} \tag{1.11}$$

can be devised.

Several disadvantages limit the effectiveness of demographic filtering. The assumption behind recommender systems based on demographic filtering is that users within the same demographic class are interested in the same items. However, in practice, demographic classes are too broad and do not catch the actual differences in the interests, preferences, and expectations of the involved users. This results in too vague recommendations. The collection of demographic data raises privacy concerns that make users reluctant to provide personal information. Finally, considerable effort is required to manually specify and periodically update the set $\mathcal{P}$ of recommendation rules.

1.3.2 CONTENT-BASED FILTERING

In content-based filtering, each item is described by a set of keywords or attributes and each user is associated with a user profile that summarizes the features of the products she liked/purchased in the past. Content-based recommendations are performed by ranking items according to a similarity function that takes into account the profile of the current user [110, 150]. We assume a scenario where items can be annotated with textual features. Hence, in its most basic form, content-based filtering is essentially an application of information retrieval techniques to the domain of recommendation.

In particular, we assume that $\mathcal{F}$ embodies a set $\{f_1, \ldots, f_q\}$ of common descriptive features for the items in $\mathcal{I}$, which summarize and index such textual content. These features can be, for example, keywords from the textual content associated with an item [14, 170], e.g., keywords extracted from the plot of the movie. Features can then be exploited to model items into a *vector representation,,* that allows us to project them into a multidimensional Euclidean space. A feature vector $\mathbf{w}_i \in \mathbb{R}^q$ is associated with an item $i \in \mathcal{I}$, and $w_{i,f}$ represents the relevance of feature f for the considered item. The weight $w_{i,f}$ can be either binary or numeric, depending on the adopted representation format[14]. For example, TF/IDF [14, 170] is an established weighting scheme that aims to penalize those features that frequently occur throughout $\mathcal{I}$, since these do not allow us to distinguish any particular item from the others.

Feature values can be exploited to predict the interest of the user for an unexperienced item. In particular, machine-learning approaches based on classification can be exploited by assuming that the previous choices of a user u in the preference matrix $\mathbf{R}$ can be partitioned into disjoint sets of positive and negative examples of her interests, i.e., $\mathcal{I}(u) = \mathcal{I}^+(u) \cup \mathcal{I}^-(u)$. For example, $\mathcal{I}^+(u)$ may consist of items with ratings above the average rating $\bar{r}_\mathbf{R}$ and, similarly, characterize $\mathcal{I}^-(u)$ as the set of items with ratings below the average. Partitioning the items allows us to build a base set upon which to train a classifier.

Besides classification, content features can be also exploited to strengthen collaborative filtering approaches, described in the next section. For example, items can be deemed similar if a suitable metric for measuring the proximity of their contents can be devised. Content similarity has been studied in several scenarios, and the most common used metrics are the following.

- **Minkowski distance** [93] generalizes the notion of distance between two points in an Euclidean space. Given the feature vectors $\mathbf{w}_i$ and $\mathbf{w}_j$ respectively associated to the items i and j, it is defined as

$$dist_p^M(i, j) = \left(\sum_{l=1}^{q} |w_{i,l} - w_{j,l}|^p \right)^{\frac{1}{p}}.$$

  Both the Euclidean and Manhattan distances originate from the above definition, for $p = 2$ and $p = 1$, respectively.

- **Cosine similarity** provides an alternative comparison for comparing items in $\mathcal{I}$. It measures the similarity of any two items in terms of the angle between their corresponding feature vectors $\mathbf{w}_i$ and $\mathbf{w}_j$:

$$sim^c(i, j) = \frac{\mathbf{w}_i^T \cdot \mathbf{w}_j}{\|\mathbf{w}_i\|_2 \cdot \|\mathbf{w}_j\|_2}.$$

  Due to its ease of interpretation and effectiveness in dealing with sparse feature vectors [52], cosine similarity is a commonly used measure.

- **Jaccard similarity** [189], originally conceived for boolean representations, tends to measure how common features tend to be predominant in the whole set of features:

$$sim^J(i, j) = \frac{\mathbf{w}_i^T \cdot \mathbf{w}_j}{\mathbf{w}_i^T \cdot \mathbf{w}_i + \mathbf{w}_j^T \cdot \mathbf{w}_j - \mathbf{w}_i^T \cdot \mathbf{w}_j}.$$

Jaccard similarity exhibits aspects of both the Euclidean and cosine measures. In particular, it tends to behave like the former (resp. the latter) if $sim^J(i, j) \to 1$ (resp. if $sim^J(i, j) \to 0$).

Choosing among the above measures is essentially a matter of how sparsely items map in the feature space. [189] provides an extensive comparison of these measures and proposes some guidelines for their adoption.

1.4 COLLABORATIVE FILTERING

State-of-the-art recommendation methods have been largely approached from a *Collaborative Filtering (CF)* perspective, which essentially consists of the posterior analysis of past interactions between users and items, aimed at identifying suitable preference patterns in users' preference data. CF techniques aim at predicting the interest of users on given items, based exclusively on previously observed preference. The assumption is that *users who adopted the same behavior in the past will tend to agree also in the future.* The main advantage of using CF techniques is their simplicity: only past ratings are used in the learning process, and no further information, like demographic data or item descriptions, is needed. This solves some of the main drawbacks of content-based and demographic approaches [114, 152].

- No personal information about users is needed; this guarantees a higher level of privacy.

- CF approaches are more general and re-usable in different contexts, while content-based techniques require the specification of a complete profile (a set of features) for each user/item.

- Content-based techniques provide the user with a list of products with features "similar" to the ones that she experienced in the past. This approach may imply the recommendations of redundant items and the lack of novelty.

- The effectiveness of recommendations increases as the user provides more feedback.

It is important to stress here that the term "collaborative" refers to the capability of summarizing the experiences on multiple users and items. There are other naive methods that still rely on the preference matrix, but they only focus on individual entries and ignore their collaborative features. Simple baselines are the item and user averages,

$$itemAvg(i) \equiv \bar{r}_i = \frac{1}{|\mathcal{U}(i)|} \sum_{u \in \mathcal{U}(i)} r_i^u \tag{1.12}$$

$$userAvg(u) \equiv \bar{r}^u = \frac{1}{|\mathcal{I}(u)|} \sum_{i \in \mathcal{I}(u)} r_i^u, \tag{1.13}$$

or more complex combinations, such as:

$$doubleCentering(u, i) = \alpha\, \bar{r}_i + (1 - \alpha)\, \bar{r}_u, \qquad (1.14)$$

where $0 \leq \alpha \leq 1$ is a parameter that can be tuned to minimize the error in prediction. More sophisticated baselines can be achieved by combining other components. In the following, we shall refer to b_i^u as any of these baseline models.

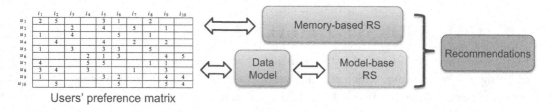

Users' preference matrix

Figure 1.2: Memory-based vs. model-based approaches.

Collaborative filtering approaches can be classified in two categories [36], namely *memory-based* and *model-based* methods. Figure 1.2 provides a sketch of the differences between them. Both categories rely on the preference matrix. However, memory-based methods infer the preference of the active user for an item by using a database of previously recorded preferences. Among memory-based approaches, a prominent role is played by neighborhood-based methods, which are based on the definition of similarity between pairs of users/items. By contrast, model-based approaches operate in two phases: in the off-line phase, the rating matrix is used to learn a compact personalization model for each user; then, the model is used in an on-line phase to predict the degree of interest of the user on candidate items.

Memory-based approaches are intuitive, as they directly transform stored preference data into predictions. The drawback is that they need access to the whole dataset to make recommendations, and, thus, they require specific indexing techniques, especially when the size of the data increases. On the other hand, model-based approaches require access to only a compact representation of the data, but the recommendation provided to the user may not be easily interpretable. A fundamental distinction also relies on the kind of relationships among users and items that they are able to exploit. Neighborhood models are effective at detecting *strong but local relationships*, as they explicitly model local similarities. Model-based approaches typically employ dimensionality reduction techniques, and hence focus on the estimation of *weak but global relationships*. Probabilistic methods, which are the focus of this manuscript, represent a refinement of the model-based approach, which relies on probabilistic modeling both in the learning phase and in the prediction phase.

In the next sections, we present a brief review of the most-used CF approaches for explicit preference data. The remainder of the manuscript analyses probabilistic methods in detail.

1.4.1 NEIGHBORHOOD-BASED APPROACHES

The idea behind *neighborhood-based approaches* [78, 174] is that similar users share common preferences, and, hence, the predicted rating on a pair $\langle u, i \rangle$ can be generated by selecting the most-similar users to u. Similar users are called *neighbors*, and their preferences on i can be aggregated and combined to produce the prediction. In a real-life scenario, this would correspond to asking friends for their opinions before purchasing a new product.

We consider here the *K-nearest-neighbors (K-NN)* approach. Within this framework, the rating prediction $\hat{r}_i^u$ is computed following simple steps: (i) a similarity function allows us to specify the degree of similarity of each a of users, thus enabling the identification of the K users most-similar to u; (ii) the rating prediction is computed as the average of the ratings given by the neighbors on the same item, weighted by the similarity coefficients. Formally, by denoting with $s_{u,v}$ the similarity between u and v and by $\mathcal{N}^K(u)$ the K most-similar neighbors of u, we have

$$\hat{r}_i^u = \frac{\sum_{v \in \mathcal{N}^K(u)} s_{u,v} \cdot r_i^v}{\sum_{v \in \mathcal{N}^K(u)} s_{u,v}}. \tag{1.15}$$

Dually, one can consider an *item-based* approach [174]: The predicted rating for the pair $\langle u, i \rangle$ can be computed by aggregating the ratings given by u on the K most-similar items to i. The underlying assumption is that the user might prefer items more similar to the ones he liked before, because they share similar features. Formally,

$$\hat{r}_i^u = \frac{\sum_{j \in \mathcal{N}^K(i;u)} s_{i,j} \cdot r_j^u}{\sum_{j \in \mathcal{N}^K(i;u)} s_{i,j}}, \tag{1.16}$$

where $s_{i,j}$ is the similarity coefficient between i and j, and $\mathcal{N}^K(i;u)$ is the set of the K items evaluated by u, which are most similar to i.

Similarity coefficients play a central role, which is two-fold: they are used to identify the neighbors, and they also act as weights in the prediction phase. In addition to the measures presented in Section 1.3.2, we can alternatively express the similarity by looking at the rating matrix. In particular, two items (resp. users) can be deemed similar if the ratings they obtained (resp. provided) are similar.

Let $\mathcal{U}_{\mathbf{R}}(i, j)$ denote the set of users who provided a rating for both i and j, i.e., $\mathcal{U}_{\mathbf{R}}(i, j) = \mathcal{U}_{\mathbf{R}}(i) \cap \mathcal{U}_{\mathbf{R}}(j)$. Two standard definitions for the similarity coefficients are the *Pearson Correlation* or the *Adjusted Cosine*. The latter is an adaptation of the cosine similarity shown in Section 1.3.2 [174]:

$$Pearson(i, j) \triangleq \frac{\sum_{u \in \mathcal{U}_{\mathbf{R}}(i,j)} \left(r_i^u - \overline{r}_i \right) \cdot \left(r_j^u - \overline{r}_j \right)}{\sqrt{\sum_{u \in \mathcal{U}_{\mathbf{R}}(i,j)} \left(r_i^u - \overline{r}_i \right)^2} \sqrt{\sum_{u \in \mathcal{U}_{\mathbf{R}}(i,j)} \left(r_j^u - \overline{r}_j \right)^2}};$$

$$AdjCosine(i, j) \triangleq \frac{\sum_{u \in \mathcal{U}_\mathbf{R}(i,j)} \left(r_i^u - \bar{r}_u\right) \cdot \left(r_j^u - \bar{r}_u\right)}{\sqrt{\sum_{u \in \mathcal{U}_\mathbf{R}(i,j)} \left(r_i^u - \bar{r}_u\right)^2} \sqrt{\sum_{u \in \mathcal{U}_\mathbf{R}(i,j)} \left(r_j^u - \bar{r}_u\right)^2}}.$$

The basic prediction Equations 1.15 and 1.16 can be extended to include unbiased adjustments. For example, the adoption of a composite baseline b_i^u allows us to tune the predictions for specific users:

$$\hat{r}_i^u = b_i^u + \frac{\sum_{j \in \mathcal{N}^K(i;u)} s_{i,j} \cdot (r_j^u - b_j^u)}{\sum_{j \in \mathcal{N}^K(i;u)} s_{i,j}}.$$

Also, it is possible to generalize the weighting scheme in the equations. In particular, within Equation 1.16, the term $s_{i,j} / \sum_{j \in \mathcal{N}^K(i;u)} s_{i,j}$ represents a weight associated with rating r_j^u, and it is fixed. We can devise a more general formulation of the prediction function, as

$$\hat{r}_i^u = \sum_{j \in \mathcal{N}^K(i;u)} w_{i,j} \cdot r_j^u. \tag{1.17}$$

Here, the term $w_{i,j}$ represents a weight associated with the pair (i, j), to be estimated. The *neighborhood relationship model* [23, 25] provides an approach to compute them as the solution of the optimization problem:

$$\min \sum_{v \in \mathcal{U}_\mathbf{R}(i)} \left(r_i^v - \sum_{j \in \mathcal{N}^K(i;u,v)} w_{i,j} \cdot r_j^v\right)^2$$

$$s.t. \quad w_{i,j} \geq 0 \sum_j w_{i,j} = 1 \; \forall i, j \in \mathcal{I}.$$

Here, the set $\mathcal{N}^K(i;u,v)$ is defined as the K items most similar to i, which are evaluated both by u and v.

The K-NN approach is characterized by a relatively lightweight learning phase: we only need to collect, for each user/item, a sufficient number of events that allow the computation of pairwise similarities. Such similarities can be directly computed online during the prediction phase. A big issue, however, is the computation of the similarity coefficients for each user/item pair. When the number of users and/or items is huge, computing such coefficients can be extremely impractical. To alleviate this problems, indexing methods are needed, which makes the search for neighbors more efficient. [205] contains a detailed survey on the subject.

1.4.2   LATENT FACTOR MODELS

The assumption behind *latent factor models* is that the overall preference of the user can be decomposed on a set of contributes that represent and weight the interaction between her tastes and the target item on a set of features. This approach has been widely adopted in information retrieval. For example, the *latent semantic indexing (LSI)* [50] is a dimensionality reduction technique that assumes a latent structure in word usage across documents. LSI employs the *singular value decomposition* to represent terms and documents in the features space: some of these feature components are very small and may be ignored, obtaining an approximate model. Given a $M \times N$ matrix $\mathbf{A}$ with rank r, the singular value decomposition of $\mathbf{A}$, denoted by $SVD(\mathbf{A})$ (see Figure 1.3), is defined as:

$$SVD(\mathbf{A}) = \mathbf{U} \times \Sigma \times \mathbf{V}^T, \tag{1.18}$$

where:

- $\mathbf{U}$ is an $M \times M$ unitary matrix; the first r columns of $\mathbf{U}$ are the eigenvectors of $\mathbf{AA}^T$ (*left singular vectors* of $\mathbf{A}$);

- $\mathbf{V}$ is an $N \times N$ unitary matrix; the first r columns of $\mathbf{V}$ are the eigenvectors of $\mathbf{A}^T\mathbf{A}$ (*right singular vectors* of $\mathbf{A}$);

- Σ is a $M \times N$ diagonal matrix with only r nonzero values, such that: $\Sigma = diag\{\sigma_1, \cdots, \sigma_n\}$, $\sigma_i \geq 0 \ \forall \ 1 \leq i < n$, $\sigma_i \geq \sigma_{i+1}$, $\sigma_j = 0 \ \forall \ j \geq r+1$;

- $\{\sigma_1, \cdots, \sigma_r\}$ are the nonnegative square root of the eigenvalues of $\mathbf{A}^T\mathbf{A}$ and are called *singular values* of $\mathbf{A}$.

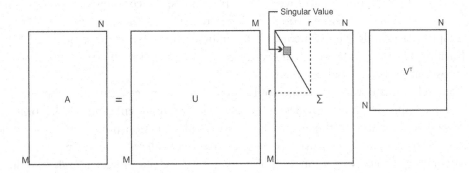

Figure 1.3: Singular value decomposition.

SVD has an important property: it provides the best low-rank linear approximation of the original matrix $\mathbf{A}$. Given a number $k \leq r$, called *dimension* of the decomposition, the matrix $\mathbf{A}_k = \sum_{i=1}^{k} u_i \sigma_i v_i^T$ minimizes the *Frobenius norm* $\|\mathbf{A} - \mathbf{A}_k\|_F$ over all rank-k matrices. Therefore, focusing only on the first k singular values of Σ and reducing the matrices $\mathbf{U}$ and $\mathbf{V}$, the original matrix can be approximated using $\mathbf{A}_k$:

$$\mathbf{A} \approx \mathbf{A}_k = \mathbf{U}_k \Sigma_k \mathbf{V}_k^T, \tag{1.19}$$

where $\mathbf{U}_k$ is obtained by removing $(M - k)$ columns from the matrix $\mathbf{U}$ and $\mathbf{V}_k^T$ is produced by removing $(N - k)$ rows from the matrix V. An example of this approximation is given in Figure 1.4. Considering the text analysis case, *LSI* factorizes the original term-document matrix

$$A \approx U_k \cdot \Sigma_k \cdot V_k^T$$

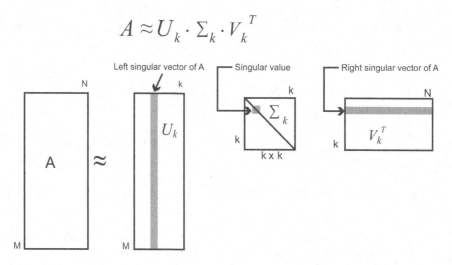

Figure 1.4: k-rank approximation of $\mathbf{A}$.

into the product of three matrices, which reflect the relationships between each single term and document in the k features-space, where k is the number of considered features. The derived matrix $\mathbf{A}_k$ is not the exact factorization of $\mathbf{A}$: the procedure of selecting only the k largest single values captures the underlying structure of $\mathbf{A}$, removing the noise at the same time [29]. Menon and Elkan in [132] provide a comparative study of the several methods for approximating the decomposition in the case of large matrices.

Several works have studied the application of SVD in recommender systems [30, 175]. A low-rank approximation provides a *low-dimensional* representation of the original *high-dimensional* rating matrix $\mathbf{R}$, thereby disclosing the hidden relationships between users and products that could be used to infer a user's preference on the considered item. If we consider a scenario involving ratings given by users on a set of movies, then we can provide a high-level interpretation of the rows of both $\mathbf{U}_k$ and $\mathbf{V}_k$. Intuitively, the row vector $\mathbf{U}_u \in \mathbb{R}^k$ maps a given user into a k-dimensional space, representing the underlying factors that influence each user's choice. By analogy, the row vector $\mathbf{V}_i \in \mathbb{R}^k$ maps a movie i into the same k-dimensional space. An example of what hidden factors might represent in this scenario is given by the movie genres. Assuming the existence of a limited number of different such genres (action, romance, comedy, etc.), the rating can be influenced by the user's preference on some genres and by the adherence of a movie to those genres. Figure 1.5 depicts this example scenario by exploiting three hidden factors.

With an abuse of notation, in this section we will use a simplified but equivalent formalization for the SVD, in which the original matrix is approximated as the product of two component

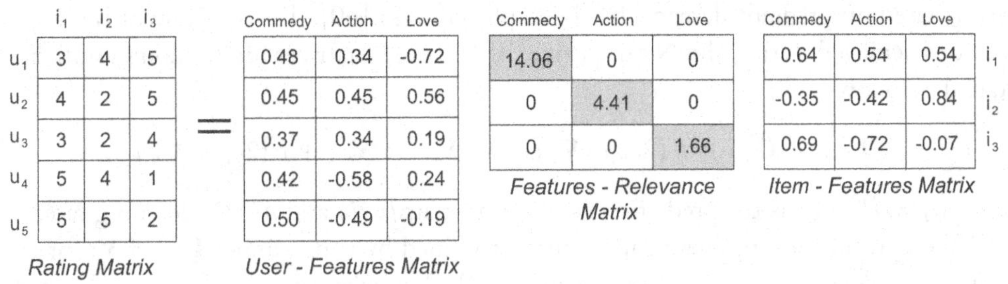

	i_1	i_2	i_3
u_1	3	4	5
u_2	4	2	5
u_3	3	2	4
u_4	5	4	1
u_5	5	5	2

Rating Matrix

Commedy	Action	Love
0.48	0.34	-0.72
0.45	0.45	0.56
0.37	0.34	0.19
0.42	-0.58	0.24
0.50	-0.49	-0.19

User - Features Matrix

Commedy	Action	Love
14.06	0	0
0	4.41	0
0	0	1.66

Features - Relevance Matrix

Commedy	Action	Love	
0.64	0.54	0.54	i_1
-0.35	-0.42	0.84	i_2
0.69	-0.72	-0.07	i_3

Item - Features Matrix

Figure 1.5: Example of the application of SVD decomposition.

matrices with K features:

$$\mathbf{R} \approx \left(\mathbf{U}_K \sqrt{\Sigma}_K^T\right)\left(\sqrt{\Sigma}_K \mathbf{V}_K^T\right) = \mathbf{U} \cdot \mathbf{V}, \tag{1.20}$$

where $\mathbf{U}$ is a $M \times K$ matrix and $\mathbf{V}$ is a $K \times N$. Intuitively, each user's preference on an item matrix is decomposed as the product of the dimensional projection of the users and items into the K-dimensional feature space:

$$\hat{r}_i^u = \sum_{k=1}^{K} \mathbf{U}_{u,k} \cdot \mathbf{V}_{k,i}. \tag{1.21}$$

The direct application of standard SVD to the context of factorizing user preferences poses some issues due to the extreme sparsity of the data. In fact, SVD requires a completely specified matrix, where all of entries are fit. In a CF scenario, missing data, i.e., unobserved user-item pairs which represent that the user did not purchase the item, can be interpreted in several ways. For instance, the user could already own the product and this purchase was not recorded on the system or he could simply not be aware of it. In other words, "the absence of evidence is not evidence of absence" and missing data requires special treatment that standard SVD does not perform. In fact, by treating as zero the preference value corresponding to unobserved user-item pairs, and applying SVD on the sparse preference matrix, the resulting model is biased toward producing low scores for items that a user has not adopted before, which may not be an accurate assumption.

To address this issue, it is convenient to minimize the prediction/reconstruction error by focusing exclusively on observed entries. Given the number of features K, and assuming that $\mathbf{U}$ and $\mathbf{V}$ represent a low-rank approximation of the original rating matrix $\mathbf{R}$, we can estimate the feature matrices by solving this optimization problem:

$$(\mathbf{U}, \mathbf{V}) = \underset{\mathbf{U}, \mathbf{V}}{\operatorname{argmin}} \left[\sum_{(u,i) \in \mathcal{T}} \left(r_i^u - \sum_{k=1}^{K} \mathbf{U}_{u,k} \mathbf{V}_{k,i} \right)^2 \right]. \tag{1.22}$$

This optimization problem has been extensively studied, both theoretically [185] and practically [60]. For example, an incremental procedure to minimize the error of the model on observed ratings, based on gradient descent, has been proposed in [60]. This was one of the major contributions achieved during the Netflix Prize. The feature matrices are randomly initialized and updated as follows:

$$\mathbf{U}'_{u,k} = \mathbf{U}_{u,k} + \eta \left(2e_{u,i} \cdot \mathbf{V}_{k,i} \right) \qquad \mathbf{V}'_{k,i} = \mathbf{V}_{k,i} + \eta \left(2e_{u,i} \cdot \mathbf{U}_{u,k} \right), \qquad (1.23)$$

where $e_{u,i} = \hat{r}^u_i - r^u_i$ is the prediction error on the pair $\langle u, i \rangle$ and η is the learning rate.

The optimization problem can be further refined by constraining $\mathbf{U}$ and $\mathbf{V}$: for example, by forcing non-negativity and sparseness [88, 112]. Particular attention has been devoted to the problem of capacity control and overfitting prevention. In a collaborative prediction setting, only some of the entries of $\mathbf{R}$ are observed. As a consequence, the generalization capabilities of the model can be compromised by the learning procedure, which can be trapped into local minima and, in particular, can be influenced by extreme values. For example, some users or items could be associated with too few observations. *Regularization* in this case aims at balancing the values of the matrices by shrinking them to more controlled values. [160, 186] suggested a formulation of this regularization termed *Maximum Margin Matrix Factorization* (MMMF). Roughly, MMMF constrains the norms of $\mathbf{U}$ and $\mathbf{V}$ to a bounded value. This corresponds to constraining the overall "strength" of the factors, rather than their number. That is, a potentially large number of factors is allowed, but only a few of them are allowed to be very important. For example, "when modeling movie ratings, there might be a very strong factor corresponding to the amount of violence in the movie, slightly weaker factors corresponding to its comic and dramatic value, and additional factors of decaying importance corresponding to more subtle features such as the magnificence of the scenery and appeal of the musical score" [160].

Mathematically, the regularization constraints can be specified within the optimization function,

$$(\mathbf{U}, \mathbf{V}) = \underset{\mathbf{U}, \mathbf{V}}{\mathrm{argmin}} \left[\sum_{(u,i) \in \mathcal{T}} \left(r^u_i - \sum_{k=1}^{K} \mathbf{U}_{u,k} \mathbf{V}_{k,i} \right)^2 + \lambda_U tr(\mathbf{U}^T \mathbf{U}) + \lambda_V tr(\mathbf{V}^T \mathbf{V}) \right], \quad (1.24)$$

where λ_U and λ_V are regularization coefficients and $tr(A)$ denotes the trace of the square matrix A. The above reformulation of the problem yields the new update rules that take into account the regularization:

$$\begin{aligned} \mathbf{U}'_{u,k} &= \mathbf{U}_{u,k} + \eta \left(2e_{u,i} \cdot \mathbf{V}_{k,i} - \lambda_U \cdot \mathbf{U}_{u,k} \right), \\ \mathbf{V}'_{k,i} &= \mathbf{V}_{k,i} + \eta \left(2e_{u,i} \cdot \mathbf{U}'_{u,k} - \lambda_V \cdot \mathbf{V}_{k,i} \right). \end{aligned} \qquad (1.25)$$

Regularization is extremely important both theoretically and practically, and has several related aspects. For example, the regularization terms can be adapted based on the popularity of users and/or items. For instance, estimation for a user who has seen very few movies will likely suffer from overfitting unless heavily regularized. Likewise, the number of ratings per movie varies widely and the regularization should take this into account. [197] further refines the optimization of Equation 1.24 by taking these aspects into account. In practice, the regularization coefficients λ_U and λ_V should be adapted to users and items, for example, by taking into account the component number of preference observation for user/item as a weighting component in the objective function. We will discuss the probabilistic interpretation of this regularization in detail in Chapter 3.

Several variations of the basic prediction rule 1.21 have been proposed. For example, the basic model can be refined by combining the baseline model with the SVD prediction [148]:

$$\hat{r}_i^u = b_i^u + \sum_{k=1}^{K} \mathbf{U}_{u,k} \cdot \mathbf{V}_{k,i}.$$

An alternative formulation reduces the number of parameters to be learned by relating the user matrix $\mathbf{U}_u$ to all the items preferred by u:

$$\mathbf{U}_{u,k} \propto \sum_{i \in \mathcal{I}(u)} \mathbf{V}_{k,i}.$$

Koren [106] further extends this model by considering both a free-factor contribution and a constrained contribution. The resulting *SVD++* model thus proposes the prediction rule:

$$\hat{r}_i^u = b_i^u + \sum_{k=1}^{K} \left(\mathbf{U}_{u,k} + \alpha \sum_{i \in \mathcal{I}(u)} \mathbf{V}_{k,i} \right) \cdot \mathbf{V}_{k,i}.$$

The gradient-descent algorithm can be tuned accordingly by plugging the above equation into Equation 1.22.

In addition to these approaches, the general task of matrix factorization for recommender systems has been widely explored and is surveyed in [107, 133]. The performance of latent factor models based on the SVD decomposition strongly depends on the number of features and the structure of the model, characterized by the presence of bias and baseline contributions. The optimization procedure used in the learning phase also plays an important role: learning can be incremental (one feature at the time) or in batch (all features are updated during the same iteration). Incremental learning usually achieves better performances at the cost of higher learning time.

1.4.3 BASELINE MODELS AND COLLABORATIVE FILTERING

We conclude this section by mentioning the relationships between the baseline methods and collaborative filtering. We discussed how the baseline methods still rely on the preference matrix, but they only focus on individual entries and ignore the collaborative nature of the data.

Besides their predictive capabilities, baselines can be exploited to refine collaborative methods, and we have seen some examples of this throughout this section. Alternatively, baselines can be used for preprocessing the data prior to the application of other techniques. The *Global Effects Model* has been introduced by *Bell and Koren* in the context of the Netflix Prize [24], and it is considered an extremely effective technique. The key idea is to adjust standard collaborative filtering algorithms with simple models that identify systematic tendencies in rating data, called *global effects*. For example, some users might tend to assign higher or lower ratings to items with respect to their average rating (*user effect*), while some items tend to receive higher rating values than others (*item effect*). Other effects can involve the time dimension: for example, temporal patterns can cause user's ratings to increase or decrease in specific periods of time, and, similarly, ratings associated with some movies can decay soon after their release. Stepwise estimation and removal of the global effects provides a substantial advantage in terms of reduction of noise. As shown in [24], adopting this pre-processing strategy enables the identification of more precise models, which are robust to the biases introduced by the global effects.

CHAPTER 2

Probabilistic Models for Collaborative Filtering

The focus of this book is the adoption of tools and techniques from probability theory for modeling users' preference data and to tackle the recommendation problem. Probability theory can be applied in several facets: for modeling past events (i.e., users' choices on a catalog of items) and making prediction about future ones; for decision theory; for model selection; etc. Clearly, many of these aspects are not specific to recommender systems, and they can be rooted instead in a more general machine-learning and statistical setting. Our focus, however, is on how the wide variety of probabilistic models can be made suitable for the recommendation data and tasks. To this purpose, next we describe both models and learning procedure and practical applications. This book is not focused on probabilistic modeling and inference by itself. Other books cover the field in a thorough way (see, e.g., [31, 138]). What we propose here is a unified view of probabilistic modeling in the context of a recommendation scenario, with particular attention to both modeling and computational aspects.

In a probabilistic setting, users' preferences are modeled as stochastic events. The analysis of the statistical distributions of such events provides insights on the mathematical rules governing users' choices and preferences. The problem then becomes how to obtain a smooth and reliable estimate of the distribution. In general, it is convenient to use a parametric model to estimate the distribution when any constraints on the shape of the distribution are known. Given a random variable X, the probability that X takes a given value x (denoted as $P(X = x)$, or just $P(x)$ for short), is parametric to a parameter set Θ, i.e., $P(x) \triangleq P(x|\Theta)$.[1] Random variables in our framework may represent either users, items, or ratings, as well as combinations of them. For example, we can model the distribution of ratings through a Gaussian distribution parameterized by $\Theta = (\mu, \sigma)$, so that $P(r) \triangleq P(r|\mu, \sigma) \sim \mathcal{N}(\mu, \sigma)$, centered on μ and with σ variance.

There are two problems we face in the following. Starting from a sample $\mathcal{X}$ of independent and identically distributed (i.i.d.) observed values $x_1, x_2, \ldots, x_n$, for a random variable X, we deal with:

- **Estimation** of Θ, that is, the specification of the set of statistical parameters based on the observed data $\mathcal{X}$.

---

[1]The notation $\triangleq$ is used in the following to denote that the term in the left-hand side is a shorthand for the term in the right-hand side.

- **Inference** of $P(\tilde{x}|\mathcal{X})$ for a given value $\tilde{x}$, based on the observed data.

The most natural and popular way to estimate parameters of the employed probabilistic model is the *maximum likelihood estimation (MLE)*, where the parameter values that are most likely to generate the observed data are chosen. That is, given a likelihood function

$$L(\Theta|\mathcal{X}) = P(x_1, x_2, \ldots, x_n|\Theta) = \prod_{i=1}^{n} P(x_i|\Theta), \tag{2.1}$$

MLE chooses the model parameter $\hat{\Theta}_{ML}$ that maximizes $L(\Theta|\mathcal{X})$, i.e.,

$$\hat{\Theta}_{ML} = \underset{\Theta}{\operatorname{argmax}} L(\Theta|\mathcal{X}).$$

The interpretation is that the desired probability distribution is the one that makes the observed data "most likely" to be observed. When $\hat{\Theta}_{ML}$ can be estimated, we can solve the inference problem as well, by approximating $P(\tilde{x}|\mathcal{X})$ as

$$P(\tilde{x}|\mathcal{X}) \approx P(\tilde{x}|\hat{\Theta}_{ML}).$$

We review some baseline methods presented in Chapter 1 and provide a probabilistic interpretation for them. First of all, we assume that the set $\mathcal{X}$ of observations is based on the rating matrix $\mathbf{R}$. Then, we can concentrate on two different scenarios: *implicit feedback* or *explicit ratings*. With implicit feedback, we assume that $\mathbf{R}$ is binary. In such cases we are interested in modeling whether, for a given pair (u, i), the property $r_i^u = 1$ holds. We can assume a Bernoulli model governed by a global parameter $\Theta \triangleq p$; thus, the probability that $r_i^u = 1$ can be modeled as $P(r_i^u = 1|\Theta) = p$. Each observation in $\mathcal{X}$ is relative to a pair (u, i) and denotes whether the associated value is positive or not. The likelihood of the data is given by

$$L(\Theta|\mathcal{X}) = \prod_{u,i} p^{r_i^u} (1 - p)^{1 - r_i^u} = p^n (1 - p)^{M \times N - n},$$

where $n = \sum_{u,i} \mathbb{1}[\![r_i^u > 0]\!]$ is the number of non-zero entries in $\mathbf{R}$. The above likelihood yields the MLE estimate

$$\hat{p}_{ML} = \frac{n}{M \times N}.$$

With explicit ratings, we still concentrate on a pair (u, i), but the value r_i^u ranges within $\{0, \ldots, V\}$. We can consider two alternative modeling strategies:

- r_i^u is sampled from a continuous distribution, e.g., a Gaussian distribution parameterized by $\Theta \triangleq \{\mu, \sigma\}$.

- r_i^u is sampled from a multinomial distribution parameterized by $\Theta \triangleq \{p_1, \ldots, p_V\}$.

In both cases, we ignore zero values, which are interpreted as unrated items. Thus, the set $\mathcal{X}$ of observed values is represented by all the pairs for which the preference value is greater than zero. Denote such pairs as $\langle u, i \rangle$: then, the likelihood can be expressed as

$$L(\Theta|\mathcal{X}) = \prod_{\langle u,i \rangle} P(r_i^u|\Theta).$$

Optimizing the above with regards to Θ yields the following:

- for the Gaussian case:

$$\hat{\mu} = \frac{1}{n} \sum_{\langle u,i \rangle} r_i^u \qquad \hat{\sigma} = \frac{1}{n} \sum_{\langle u,i \rangle} (r_i^u - \hat{\mu})^2,$$

- for the multinomial case:

$$p_r = \frac{1}{n} \sum_{\langle u,i \rangle} \mathbb{1}[\![r_i^u = r]\!].$$

Throughout the chapter, we shall use graphical models representing Bayesian networks to denote variables and dependencies among elements [101]. Graphical models express the joint distribution of a system or phenomenon in terms of random variables and their conditional dependencies in a *Directed Acycling Graph (DAG)*. Nodes of this DAG represent random variables, while edges correspond to causal dependencies that entail conditional probability distributions. Bayesian networks distinguish between evidence nodes (pictured with a gray background in the following), which correspond to variables that are observed or assumed observed, and hidden nodes, which correspond to latent variables. Many models are specified by replications of nodes that share parents and/or children, e.g., to account for multiple observations from the same distribution or mixture components. Such replications can be denoted by plates, which surround the subset of nodes and have a replication count or a set declaration of the index variable at the lower right corner. Graphical models provide an intuitive description of an observed phenomenon as a so-called generative model, which states how the observations could have been generated.

Consider the following graphical model:

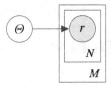

This DAG specifies a causal relationships between the latent random variable Θ and the observed data r. The two plates respectively specify users (M) and items (N). A value r is associated with each combination (u, i) (expressed by the intersection of the boxes).

We can observe that the resulting estimations provide a probabilistic interpretation of the OverallMean baseline method discussed in Chapter 1: in such a model we are interested in devising a global behavior from the rating matrix $\mathbf{R}$, upon which to base our predictions. However, we can consider more refined situations, where we can condition the observation r to the identity of the user/item involved. For example, we can specialize the above estimations to resemble the UserAvg method. This model assumes that the parameter set is specific to each user, i.e., $\Theta \triangleq \{\Theta_1, \ldots, \Theta_M\}$, such that $P(r_i^u|\Theta) \triangleq P(r_i^u|\Theta_u)$.

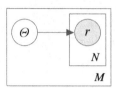

Hence, $\mathcal{X}$ represents the rows of $\mathbf{R}$, where each row comprises all the possible observations involving a user u. In the case of implicit preferences, Θ_u represents a Bernoulli parameter p_u, and we can rewrite the likelihood as

$$L(\Theta|\mathcal{X}) = \prod_u \prod_i p_u^{r_i^u} (1 - p_u)^{1-r_i^u} = \prod_u p_u^{n_u}(1 - p_u)^{M-n_u},$$

which can be optimized with respect to each p_u separately, yielding $\hat{p}_u = n_u/M$, where $n_u = |\mathcal{I}(u)|$.

In the rest of the chapter we study probabilistic modeling by following an incremental approach. We start in Section 2.1, by reformulating the rating prediction problem in a probabilistic setting and we propose a simple probabilistic framework that reconsiders and strengthens the naive Gaussian and multinomial models shown, so far, in a regression perspective. We then introduce latent factor modeling in Section 2.2, and gradually show how the latter can be exploited to model more complex relationships which hold within preference data. It is worth noting that most of the models described in the next sections are extremely important from a theoretical point of view, as they represent simple yet mathematically robust alternative research directions for modeling preferences. However, they should only be considered under a pedagogical perspective, since their predictive abilities were historically overcome by their reformulation in a Bayesian setting, which we shall cover in details in Chapters 3 and 5.

2.1   PREDICTIVE MODELING

We assume three random variables X, Y, and R, ranging over $\mathcal{U}, \mathcal{I}$, and $\mathcal{V}$. The event $X = u, Y = i, R = r$ represents the fact that $\mathbf{R}$ exhibits the value r in correspondence to row u and column i. This allows three possible modeling alternatives for prediction modeling: either, $P(r|u, i)$, $P(r, i|u)$, or $P(r, u, i)$. Although there is strong relationship between these quantities, the underlying models are different and correspond to different scenarios. The first setting, called *forced prediction* [85], involves predicting a preference value for a particular item i, given the identity of the user. In such a context, we are interested in the probability $P(r|u,i)$ that user u will rate item i with r. Based on the conditional probability, we can define a prediction function $g(u,i)$, which depends on the probability $P(r|u,i)$: for example, by associating $g(u,i)$ to a deterministic value $g(u,i) = \text{argmax}_r \, P(r|u,i)$ or to the mean value $g(u,i) = \int_r r \cdot P(r|u,i) \, dr$. Hoffman denotes this setting forced prediction because it mimics an experimental setup in which the response of a user is solicited for a particular item and the user has no choice on which item to vote. This is the relevant prediction mode in scenarios in which an item is presented to a user as a recommendation and one is interested in predicting the user's response.

The second choice, i.e., $P(r,i|u)$, is called *free prediction* and it explains not only the observed ratings, but also why item i is chosen to be rated by user u. By using the chain rule, we have $P(r,i|u) = P(r|u,i) \cdot P(i|u)$, thus decomposing the problem into the prediction of the selection of the considered item (irrespective of the rating) and a prediction of the rating conditioned on the (hypothetically) selected item. The probability $P(i|u)$ corresponds to the likelihood of an implicit preference, whereas $P(r|u,i)$ corresponds to the probability of an explicit preference. Thus, free prediction can incorporate both models and it achieves a more comprehensive modeling of the preference phenomena. Moreover, Hoffman [85] highlights that the free prediction case is a generalization of what is commonly referred to as the "recommendation" task, i.e., selecting a set of items to present to the user. As a result, items that have been rated by many users will have more impact on the model estimation than items that are only rated by a few users.

The third choice, i.e., $P(r,u,i)$, models the joint distribution between the three random variables. Under this choice, the model is also concerned with the behavior of users (e.g., some users tend to purchase and rate more items than the average). In particular, users with more ratings tend to have a larger impact on the final model than users that only rate a few items. By using the chain rule, $P(r,u,i) = P(r|u,i)P(i|u)P(u)$ and we can notice the role played by the prior probability about user u.

Let us concentrate on forced prediction for the moment. The objective is to model the probability $P(r|u,i)$ that user u will rate item i with rating r. The pair (u,i) can be characterized by a set of *features*: these are elementary pieces of evidence that link aspects of what we observe relative to (u,i) with the rating r that we want to predict. Example features can be indicator functions of the atomic properties in $\mathcal{F}$ described in Chapter 1, such as Sex(u,male) or Rating(i,PG), or even their combination, such as Sex(u,male) $\wedge$ Rating(i,PG). In its most general form, a feature can be considered as a function $f : \mathcal{U} \times \mathcal{I} \mapsto \mathbb{R}$ associating a real value to an observation $\langle u,i \rangle$. In the above example, a relative feature can be encoded as $\mathbb{1}[\![\text{Sex(u,male)} \wedge \text{Rating(i,PG)}]\!]$.

Features are important because they map an observation $\langle u,i,r \rangle$ to a multidimensional numerical space. By selecting a fixed set $\mathcal{F} = \{f_1, \ldots, f_d\}$ of features, each observation $\langle u,i \rangle$ can be encoded as a d-dimensional vector, and then we can approach the prediction problem as a regression problem. In particular, by interpreting r as a numerical value, we can introduce a set of parameters $\Theta \triangleq \{\beta_1, \ldots, \beta_d, \sigma\}$, and model the probability $P(r|u,i)$ by means of a *Gaussian model*:

$$P(r|u,i,\Theta) = \frac{1}{\sqrt{2\pi}\sigma} \exp\left\{-\frac{1}{2\sigma^2}(r-\mu_{u,i})^2\right\}, \tag{2.2}$$

where

$$\mu_{u,i} = \sum_{k=1}^{d} \beta_k f_k(u,i).$$

Analogously, by interpreting r as a discrete value, we can refine the simple multinomial model proposed in the beginning of this chapter, by assuming that the multinomial parameter p_r is specific to the observation $\langle u,i,r\rangle$ at hand. Consider a set of logistic factors $\Theta \triangleq \{\beta_{r,1}, \ldots, \beta_{r,d}\}$. Then we can define the probability of observing r relative to $\langle u,i\rangle$ by means of the *log-linear* model

$$P(r|u,i,\Theta) = \frac{1}{Z_{u,i}(\Theta)} \exp\left\{\sum_k \beta_{r,k} f_k(u,i)\right\}, \tag{2.3}$$

where $Z_{u,i}(\Theta)$ is a normalization factor, expressed as

$$Z_{u,i}(\Theta) = \sum_{r\in\mathcal{V}} \exp\left\{\sum_k \beta_{r,k} f_k(u,i)\right\}.$$

This simple yet powerful modeling has been studied extensively in the literature [28, 131, 149, 212].[2] In its most basic form, the model can be developed based on elementary features, relative to the dimensions u and i of the preference [131]: $f_{\langle u\rangle}$ and $f_{\langle i\rangle}$ (where $\langle\cdot\rangle$ represents an enumeration of features relative to users/items), can be defined as simple indicator functions,

$$f_{\langle u'\rangle}(u,i) = \mathbb{1}[\![u = u']\!] \qquad f_{\langle i'\rangle}(u,i) = \mathbb{1}[\![i = i']\!],$$

and we can assume $M \times N$ features, which enable the corresponding regression factors. Figure 2.1 shows a toy preference matrix and the set of features associated with each observation. Some patterns are clearly observable within the feature matrix: for example, f_4 and f_2 are associated with low preference values.

Besides this naive modeling, it is possible to include even more complex features, representing, e.g., side information or dependency among users/items (see Chapter 6).

The estimation of the weights associated with the features can be accomplished by means of MLE, by optimizing the log-likelihood $LL(\Theta|\mathcal{X}) = \sum_{\langle u,i,r\rangle} \log P(r|u,i,\Theta)$. In the Gaussian

---

[2]Log-linear models are presented in a more general notation in the current literature by assuming that feature functions also embody the value to predict: that is, a generic $f_k : \mathcal{U} \times \mathcal{I} \times \mathcal{V} \mapsto \mathbb{R}$ is relative to the tuple (u,i,r). The notation we propose here is a particular case of this more general formulation, and it is simplified for highlighting the analogy with the Gaussian model.

	f_1	f_2	f_3	f_4	f_5	f_6	f_7
$\langle u_1,i_1,1\rangle$	1			1			
$\langle u_1,i_3,1\rangle$			1	1			
$\langle u_2,i_3,2\rangle$			1			1	
$\langle u_2,i_4,2\rangle$		1					1
$\langle u_3,i_1,5\rangle$	1				1		
$\langle u_3,i_3,4\rangle$		1				1	

	i_1	i_2	i_3	i_4
u_1	1	5		
u_2			2	2
u_3	1		4	

(a) Users' preference matrix   (b) Features associated with observations

Figure 2.1: Users' preference matrix and features.

case, we have

$$LL(\Theta|\mathcal{X}) = -\frac{n}{2}\log\sigma - \frac{n\log\pi}{2} - \frac{1}{2\sigma^2}\sum_{\langle u,i,r\rangle}\left(r - \sum_k \beta_k f_k(u,i)\right)^2, \quad (2.4)$$

which yields the classical solution

$$\hat{\boldsymbol{\beta}} = (\boldsymbol{\Phi}^T\boldsymbol{\Phi})^{-1}\boldsymbol{\Phi}^T\mathbf{r}, \quad (2.5)$$

$$\hat{\sigma} = \frac{1}{n}\sum_{\langle u,i,r\rangle}\left(r - \sum_k \hat{\beta}_k f_k(u,i)\right)^2, \quad (2.6)$$

where $\boldsymbol{\Phi}$ is the $m \times d$ *design* matrix relative to all the features associated with all the observations in $\mathcal{X}$, and $\mathbf{r}$ is the vector of all ratings associated with such observations.

For the discrete case, consider the log-likelihood

$$LL(\Theta|\mathcal{X}) = \sum_{\langle u,i,r\rangle}\sum_k \beta_{r,k} f_k(u,i) - \sum_{u,i}\log\sum_{r\in\mathcal{V}}\exp\left\{\sum_k \beta_{r,k} f_k(u,i)\right\}.$$

Unfortunately, the summation inside the logarithm does not provide any closed formula, for the $\beta_{r,k}$ parameters, so one has to resort to iterative methods in order to approximate the optimal solution. Historically, the optimization problem has been approached by means of *iterative scaling* techniques [48]. These methods are specialized on log-linear models: they iteratively construct a lower bound to the log-likelihood, and then optimize the bound. People have worked on many variants of the bounding technique, and an efficient variant is the *Improved Iterative Scaling (IIS)* algorithm. Starting with an arbitrary solution Θ, IIS iteratively searches for an improvement $\delta_{r,k}$ for each $\beta_{r,k}$, such that $\Theta + \Delta$ now represents the parameters $\beta_{r,k} + \delta_{r,k}$, and this new parameter set guarantees $LL(\Theta + \Delta|\mathcal{X}) \geq LL(\Theta|\mathcal{X})$. We can observe the following:

$$LL(\Theta + \Delta|\mathcal{X}) - LL(\Theta|\mathcal{X})$$

$$= \sum_{\langle u,i,r \rangle} \sum_k (\beta_{r,k} + \delta_{r,k}) f_k(u,i) - \sum_{u,i} \log \sum_{r \in \mathcal{V}} \exp \left\{ \sum_k (\beta_{r,k} + \delta_k) f_k(u,i) \right\}$$

$$- \sum_{\langle u,i,r \rangle} \sum_k \beta_{r,k} f_k(u,i) + \sum_{u,i} \log \sum_{r \in \mathcal{V}} \exp \left\{ \sum_k \beta_{r,k} f_k(u,i) \right\}$$

$$= \sum_{\langle u,i,r \rangle} \sum_k \delta_{r,k} f_k(u,i) - \sum_{u,i} \log \frac{\sum_{r \in \mathcal{V}} \exp \left\{ \sum_k (\beta_{r,k} + \delta_{r,k}) f_k(u,i) \right\}}{\sum_{r \in \mathcal{V}} \exp \left\{ \sum_k \beta_{r,k} f_k(u,i) \right\}}.$$

We can apply the property $\log x \leq x - 1$, to obtain

$$LL(\Theta + \Delta|\mathcal{X}) - LL(\Theta|\mathcal{X})$$

$$\geq \sum_{\langle u,i,r \rangle} \sum_k \delta_{r,k} f_k(u,i) - \sum_{u,i,r} \frac{\exp \left\{ \sum_k (\beta_{r,k} + \delta_{r,k}) f_k(u,i) \right\}}{\sum_{r \in \mathcal{V}} \exp \left\{ \sum_k \beta_{r,k} f_k(u,i) \right\}}$$

$$= \sum_{\langle u,i,r \rangle} \sum_k \delta_{r,k} f_k(u,i) - \sum_{u,i,r} P(r|u,i,\Theta) \exp \left\{ \sum_k \delta_{r,k} f_k(u,i) \right\}.$$

The above formula can be further simplified by exploiting Jensen's inequality[3]. By introducing $f^{\#}(u,i) = \sum_k f_k(u,i)$, we have:

$$\sum_{\langle u,i,r \rangle} \sum_k \delta_{r,k} f_k(u,i) - \sum_{u,i,r} P(r|u,i,\Theta) \exp \left\{ \sum_k \delta_{r,k} f_k(u,i) \right\}$$

$$= \sum_{\langle u,i,r \rangle} \sum_k \delta_{r,k} f_k(u,i) - \sum_{u,i,r} P(r|u,i,\Theta) \exp \left\{ f^{\#}(u,i) \sum_k \delta_{r,k} \frac{f_k(u,i)}{f^{\#}(u,i)} \right\}$$

$$\geq \sum_{\langle u,i,r \rangle} \sum_k \delta_{r,k} f_k(u,i) - \sum_{u,i,r} P(r|u,i,\Theta) \sum_k \frac{f_k(u,i)}{f^{\#}(u,i)} \exp \left\{ f^{\#}(u,i) \delta_{r,k} \right\}$$

$$\triangleq \mathcal{B}(\Theta, \Delta).$$

Thus, $\mathcal{B}(\Theta, \Delta)$ represents a lower bound of the difference $LL(\Theta + \Delta|\mathcal{X}) - LL(\Theta|\mathcal{X})$. Hence, by optimizing $\mathcal{B}(\Theta, \Delta)$ with respect to Δ, we can guarantee an improvement of the likelihood. A nice property of $\mathcal{B}(\Theta, \Delta)$ is that all the $\delta_{r,k}$ parameters are only bound by linear relationships, so that optimizing $\mathcal{B}(\Theta, \Delta)$ separately for each $\delta_{r,k}$ is relatively easy. As a matter of fact, the optimization of $\mathcal{B}(\Theta, \Delta)$ with respect to $\delta_{r,k}$ admits a closed formula. The overall IIS procedure can be finally schematized as follows.

[3]The inequality states that, given a convex function f and a random variable X ranging over the domain of f, then $E[f(X)] \geq f(E[X])$. Details can be found in [45, Section 2.6].

- Start with an initial arbitrary choice for each $\beta_{r,k}^0$;

- Repeat until convergence for increasing steps t:

    1. Optimize $\mathcal{B}(\Theta^{(t)}, \Delta)$ with respect to $\delta_{r,k}$, for each r and k;

    2. Update $\beta_{r,k}^{(t+1)} = \beta_{r,k}^{(t)} + \delta_{r,k}$.

Iterative scaling methods have three main advantages. They can easily incorporate feature selection; furthermore, they scale up well in the number of features; finally feature dependencies do not affect the accuracy of the learned model. The biggest problem with these methods is that the bound provided at each iteration can be loose. As a consequence, the rate of convergence can be extremely slow. Improvements to the basic scaling scheme have been proposed [89, 99], which provide tighter bounds and in general tend to increase the rate of convergence.

It is generally recognized that gradient-based optimization algorithms [144] are able to find the optimal parameters much faster than the iterative scaling methods. Among these, general-purpose methods that use the Newton search direction have a fast rate of local convergence, typically quadratic. The main drawback of these methods is the need for the Hessian matrix of the log-likelihood. Explicit computation and inversion of this matrix can be problematic, when the number of feature functions is high. Quasi-Newton methods provide a practical alternative, in that they rely on an approximation of the Hessian, which can be computed incrementally, and still attain a superlinear rate of convergence. Limited-memory quasi-Newton methods, like L-BFGS [116], represent the current methods of choice for training log-linear models [120].

2.2   MIXTURE MEMBERSHIP MODELS

The exceptional sparsity of users' preference data poses a main challenge for users' profiling and the generation of personalized recommendations. Clustering and, in general, dimensionality reduction techniques, allow us to abstract from the limited available data for a given user or item, and identify preference patterns that can be associated with each user or item with a certain degree of certainty. This is the underlying idea of mixture membership models: analyzing and detecting similarities among users and/or items and introducing a set of latent variables, which identify a fixed number of more general preference patterns. Generally, the design of such models is driven by the following considerations:

- **Detection of similarity among users or items.** In other words, given observed preference data, how should we model and detect user or item similarity? For instance, similar-minded users can be detected simply by considering the number of common purchased items, or by considering their agreement on ratings. More complex similarity measures might highlight the mutual relationships between similar users and similar items.

- **Choice of causal relationships to be modeled.** While entities to be modeled in the context of preference data are fixed, i.e., user, items, preference values, and, possibly, latent variables, their causal relationships may vary. In a user-based modeling scenario, we are interested in abstracting from the level of single users and identifying preference patterns that are well-suited for a group of similar minded users. Symmetrically, in an item-based perspective, the goal is to identify categories of items that tend to receive similar evaluations from users. These dependencies can be combined into more sophisticated models, which allow simultaneous clustering on both dimensions and hierarchical relationships.

- **Choice of the membership.** Should each user/item be statically associated with a given cluster, or should we allow fuzzy memberships? In the first case, the model induces a partition over the user set or the item set by identifying user communities or item categories. In the second case, we assume that user/item can have several interests/aspects and, hence, they can be associated with different clusters simultaneously.

- **The modeling of preference patters, i.e., the choice of the appropriate generative distribution for preference values.** Several choices can be taken to model rating patterns: if ratings are discrete, as in the case of the typical 5 stars scale, a multinomial distribution can be employed to model the probability of observing a given rating value. If ratings are continuous, the choice of a Gaussian distribution is more appropriate.

This section is aimed at reconsidering the basic models proposed in the beginning of this chapter under a clustering perspective, according to the aforementioned dimensions.

In a nutshell, clustering can be interpreted in terms of a latent causal relationship that influences a user's choice or an item characteristic. Hence, latent factors can be exploited to model similarities and to express the likelihood to observe a preference value in terms of such similarities. The simple probabilistic models, proposed at the beginning of the chapter, can be specialized to cover the case where a user's choice is influenced by a latent factor. That is, a generic event x can be conditioned to a variable y representing a latent state, so that $P(x)$ can be expressed as a mixture

$$P(x) = \sum_y P(x|y)P(y),$$

and we can concentrate on modeling both $P(x|y)$ and $P(y)$. More generally, a *mixture model* [31] is a general probabilistic approach that introduces a set of latent factor variables to represent a generative process in which data points may be generated according to a finite set of probability distributions. The introduction of the latent variables allows the modeling of complex distributions in terms of simpler components.

A naive way of introducing mixtures within preference data is to consider that each user is associated with a latent state, and, consequently, to model the probability $P(r_i^u)$ accordingly. We can consider a $M \times K$ binary matrix $\mathbf{Z}$, where each row $\mathbf{z}_u$ is such that $\sum_k z_{u,k} = 1$. We will also use the single multinomial latent variable z_u (or z, when u is clear from the context) with values ranging in $\{1, \ldots, K\}$, with $z_u = k$ denoting the fact that $z_{u,k} = 1$.

The vector $\mathbf{z}_u$ is associated with a user u, and, in particular, with a row $\mathbf{r}_u$ of $\mathbf{R}$. Intuitively, $\mathbf{z}_u$ represents the fact that user u is associated with the one factor k, such that $z_{u,k} = 1$, and, hence, all its preferences can be explained in terms of k. That is, the parameter set Θ can be decomposed into $\Theta_1, \ldots, \Theta_K$, and the preferences of u are generated according to Θ_k. Graphically, the generative process is expressed in Figure 2.2.

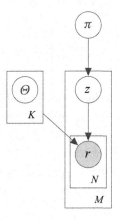

Figure 2.2: Graphical model for the mixture model.

Here, $\boldsymbol{\pi}$ is a multinomial distribution that models the prior probability $P(z|\boldsymbol{\pi})$ relative to a latent factor z. The underlying idea is that data are generated according to a stochastic process governed by the underlying probability distributions. This process can be expressed as follows:

- For each user u pick a latent factor $k \sim Disc(\boldsymbol{\pi})$;
  - For each item i pick a preference $r \sim P(r|\Theta_k)$.

According to this model, the probability of observing $\mathbf{r}_u$ is given by

$$P(\mathbf{r}_u|\Theta, \boldsymbol{\pi}) = \sum_z P(z|\boldsymbol{\pi})P(\mathbf{r}_u|z, \Theta) = \sum_{k=1}^{K} \pi_k \prod_{i:\langle u,i\rangle \in \mathbf{R}} P(r_i^u|\Theta_k),$$

and consequently the likelihood can be expressed as

$$L(\Theta, \boldsymbol{\pi}|\mathcal{X}) = \prod_u \sum_{k=1}^{K} \pi_k \prod_{\langle u,i\rangle} P(r_i^u|\Theta_k). \tag{2.7}$$

The optimization of the above likelihood is difficult, and one has resort to iterative techniques, based on numerical optimization. An alternative approach is given by resorting to the EM algorithm [51]. Similar to the IIS scheme discussed before, the EM algorithm is based on the specification of a lower bound to the true likelihood, which hence can be exploited within an iterative scheme. The EM approach (see Appendix A.1 for the mathematical basis of the algorithm) consists of considering the complete-data expectation log-likelihood:

$$
\mathcal{Q}(\Theta, \Theta') = \sum_{\mathbf{Z}} P(\mathbf{Z}|\mathcal{X}, \Theta') LL(\Theta|\mathcal{X}, \mathbf{Z})
$$

$$
= \sum_{u} \sum_{k=1}^{K} \gamma_{u,k}(\Theta') \left\{ \log \pi_k + \sum_{\langle u,i \rangle} \log P(r_i^u|\Theta_k) \right\},
$$

where the term $\gamma_{u,k}(\Theta')$ represents the posterior probability of the latent variable $z_{u,k}$ given Θ' and u. Hence, the core of the algorithm is an iterative alternation between the *expectation* (E) step, where we infer the term $\gamma_{u,k}$, and the *maximization* (M) step, where we exploit the $\gamma_{u,k}$ computed so far to optimize $\mathcal{Q}(\Theta, \Theta')$ with respect to Θ.

Notably, by properly instantiating $P(r_i^u|\Theta_k)$, we can obtain a complete specification of both the E and the M steps.[4] Let us review the models proposed at the beginning of this chapter, within a mixture-modeling framework.

Example 2.1 (Bernoulli model)   Events in this model are represented by the presence/absence of a preference. Define $\Theta_k \triangleq p_k$, $P(r|\Theta_k) = p_k$ and let n_u denote the number of items previously purchased by the user u, i.e., $n_u = |\mathcal{I}(u)|$. Then,

$$
\mathcal{Q}(\Theta, \Theta') = \sum_{u} \sum_{k=1}^{K} \gamma_{u,k}(\Theta') \left\{ \log \pi_k + n_u \cdot \log p_k + (M - n_u) \cdot \log(1 - p_k) \right\},
$$

which defines the M-step as:

$$
\hat{p}_k = \frac{\sum_u \gamma_{u,k} \cdot n_u}{M \sum_u \gamma_{u,k}}.
$$

Example 2.2 (Explicit preferences)   Adapting the framework is straightforward, as it just requires re-defining $\Theta_k \triangleq \{\mu_k, \sigma_k\}$ (for the continuos/Gaussian case), or $\Theta_k \triangleq \{p_{k,1}, \dots, p_{k,V}\}$ (for the discrete case). Since the observations $\mathcal{X}$ are only relative to the rated items, we have:

---

[4]In particular, the term $\gamma_{u,k}$ can be inferred by means of Equation A.7, independently of the functional form of $P(r_i^u|\Theta_k)$.

- for the Gaussian case,

$$\mathcal{Q}(\Theta', \Theta') \propto \sum_u \sum_{k=1}^{K} \gamma_{u,k}(\Theta') \left\{ \log \pi_k - n_u \log \sigma_k - \sigma_k^{-2} \sum_{i:r_i^u > 0} (r_i^u - \mu_k)^2 \right\},$$

and

$$\hat{\mu}_k = \frac{\sum_{u,i} \gamma_{u,k} \cdot r_i^u}{\sum_u \gamma_{u,k} \cdot n_u} \qquad \hat{\sigma}_k = \frac{\sum_{\langle u,i \rangle:r_i^u > 0} \gamma_{u,k} \cdot (r_i^u - \hat{\mu}_k)^2}{\sum_u \gamma_{u,k} \cdot n_u};$$

- for the multinomial case,

$$\mathcal{Q}(\Theta', \Theta') = \sum_u \sum_{k=1}^{K} \gamma_{u,k}(\Theta') \left\{ \log \pi_k + \sum_{i:r_i^u > 0} \sum_r \mathbb{1}[\![r_i^u = r]\!] \log p_{k,r} \right\},$$

with $\sum_r p_{k,r} = 1$, and

$$p_{k,r} = \frac{\sum_u \gamma_{u,k} \cdot \sum_i \mathbb{1}[\![r_i^u = r]\!]}{\sum_u \gamma_{u,k} \cdot n_u}.$$

2.2.1   MIXTURES AND PREDICTIVE MODELING

Exploiting a mixture model for prediction is straightforward. For example, in the case of forced prediction, we can decompose the probability distribution over ratings for the pair $\langle u, i \rangle$ as

$$P(r|i, u, \Theta) = \sum_z P(z|u, i, \Theta) \cdot P(r|z, \Theta).$$

Observe now that, in the models previously introduced, the latent factor is only associated with a user, that is, $P(z|u, i, \Theta) = P(z|u, \Theta)$, which we can approximate by using the posterior probability $\gamma_{u,k}(\Theta)$, given the observations in $\mathcal{X}$. Hence, a rating distribution can be obtained by instantiating $P(r|z, \Theta)$ with any of the models discussed so far.

This naive model can be further refined in several ways. First of all, observe that, in this model, the probability of a rating is governed by the component $P(r|z, \Theta)$, which expresses a general trend relative to the ratings generated by all the users associated with a given latent factor z. As a consequence, the model is not able to capture finer distinctions that take into account items. Consider the following example of a preference matrix.

1	3		5	2	3	4	
	2	1		1	3		1
5		1	2		2		
		2	1		1	2	2
	1	2		3		2	

Two clusters can be devised, which group users u_1, u_2, and u_3, u_4, u_5. In particular, within the second group, a general tendency is to provide low preference values, except for item i_i, for which both u_3 and u_4 provide a high preference. This observation highlights two different preference patterns within the second cluster: high values for item i_1, and low values for the remaining items. However, these two opposing patterns are not caught by either the multinomial nor the Gaussian mixture model defined in the previous section, as their parameters tend to average the ratings according to the general trend.

To avoid such a drawback, it is convenient to model the prediction probability by enforcing the dependency of the rating on both the latent factor and the proposed item [124] , as

$$P(r|i, u, \Theta) = \sum_z P(z|u, \Theta) \cdot P(r|z, i, \Theta).$$

Figure 2.3(a) adapts the simple multinomial mixture model to such a situation. Here, the latent factor z is associated with a specific user and the rating is given by a multinomial probability distribution $\epsilon_{i,k}$, which depends on both the status k of the latent variable and the user i under observation. As a consequence, the likelihood relative to the parameter set $\Theta = \{\pi, \epsilon\}$ can be expressed as

$$P(\mathcal{X}|\Theta) = \prod_u \sum_k \pi_k \prod_{\langle u,i,r \rangle} P(r|\epsilon_{i,k}).$$

The adaptation of procedure for learning ϵ, as well as the formulation of the Gaussian case, are straightforward.

The *User Communities Model* (UCM, [21]) provides a further refinement, by focusing on free prediction and thus modeling the probability $P(r, i|u)$. The underlying generative process is shown in Figure 2.3(b) and can be expressed as follows.

- For each user u:

    - Pick a latent factor $z \sim Disc(\pi)$;
    - For $j = 1$ to n_u:

        select an item $i \sim P(i|\phi_z)$;
        generate a preference value of the selected item $r \sim P(r|\epsilon_{i,z})$.

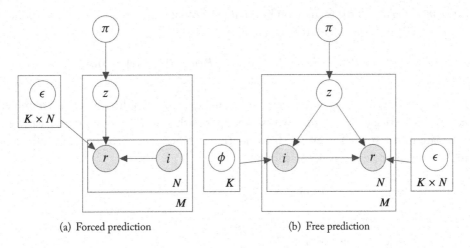

(a) Forced prediction   (b) Free prediction

Figure 2.3: Graphical model for the mixture model.

Here, ϕ_z denotes a multinomial probability $\{\phi_{z,i}\}_{i \in \mathcal{I}}$ among items, and $\epsilon_{i,z}$ the parameter set for the rating, relative to latent factor z when the observed item is i. $P(r|\epsilon_{i,k})$ can be instantiated either as a multinomial or a Gaussian distribution. The corresponding likelihood can be expressed as

$$P(\mathcal{X}|\Theta) = \prod_u \sum_k \pi_k \prod_i \prod_{r=1}^{V} \{\phi_{k,i} P(r|\epsilon_{i,k})\}^{\mathbb{1}[r_i^u > 0]}.$$

The EM framework provides an estimation of the model parameters. Predicting the likelihood of a preference i, r for a given user u is straightforward:

$$P(r, i|u, \Theta) = \sum_z P(z|u, \Theta) \cdot P(i, r|z, \Theta)$$
$$= \sum_k \gamma_{u,k} \phi_{k,i} P(r|\epsilon_{i,k}),$$

where we exploit the responsibility $\gamma_{u,k}$ representing the posterior probability of observing the factor k given user u and the observations $\mathcal{X}$.

As of now, we associated a latent variable to a user. Intuitively, this allows us to group users in clusters of individuals with similar preferences, expressed by either similar ratings or similar selections. It is also possible to model the opposite situation, where a latent variable allows us to group items that received similar preferences. However, latent variables can be used to model more complex situations, where both a user and an item can be associated with the same latent variable, and all the components of an observation can be thought of as the result of a stochastic process.

The *aspect model* (AM, [86]) focuses on implicit preferences and explains both components of an observation $\langle u, i \rangle$ in terms of a latent variable:

$$P(u, i | \Theta) = \sum_z P(z | \boldsymbol{\pi}) \cdot P(u | z, \boldsymbol{\theta}) \cdot P(i | z, \boldsymbol{\phi}).$$

Here, $P(u | z, \boldsymbol{\phi})$ and $P(i | z, \boldsymbol{\phi})$ are class-conditional multinomial distributions parameterized by $\boldsymbol{\theta}_z$ and $\boldsymbol{\phi}_z$, respectively. In practice, we assume that both a user u and an item i can be selected to participate in the stochastic event. The underlying generative process, graphically depicted in Figure 2.4(a), can be hence devised as follows.

- For each rating observation n:

  – Pick a latent factor $z \sim Disc(\boldsymbol{\pi})$;
    select a user $u \sim Disc(\boldsymbol{\theta}_z)$;
    select an item $i \sim Disc(i | \boldsymbol{\phi}_z)$.

The data likelihood can be accordingly expressed as

$$P(\mathcal{X} | \Theta) = \prod_{\langle u, i \rangle} P(u, i | \Theta) = \prod_{\langle u, i \rangle} \sum_z P(z | \boldsymbol{\pi}) \cdot P(u | z, \boldsymbol{\theta}) \cdot P(i | z, \boldsymbol{\phi}).$$

By applying the EM framework, the complete data expectation log-likelihood can be expressed as

$$\mathcal{Q}(\Theta, \Theta') = \sum_{\langle u, i \rangle} \sum_k \gamma_{u,i,k}(\Theta') \left\{ \log \pi_k + \log \theta_{u,k} + \log \phi_{i,k} \right\},$$

which finally results in the update equations

$$\gamma_{u,i,k} = \frac{\pi_k \cdot \theta_{u,k} \cdot \phi_{i,k}}{\sum_{j=1}^K \pi_j \cdot \theta_{u,j} \cdot \phi_{i,j}}, \qquad \theta_{u,k} = \frac{\sum_{i : r_i^u > 0} \gamma_{u,i,k}}{\sum_{u', i : r_i^{u'} > 0} \gamma_{u',i,k}}, \qquad \phi_{i,k} = \frac{\sum_{u : r_i^u > 0} \gamma_{u,i,k}}{\sum_{u, i' : r_{i'}^u > 0} \gamma_{u,i',k}}.$$

The adaptation of the aspect model to incorporate rating information is trivial and it can be done in two different ways: either by assuming that the rating is independent from the item,

$$P(u, i, r | \Theta) = \sum_z P(z | \boldsymbol{\pi}) \cdot P(u | z, \boldsymbol{\theta}) \cdot P(i | z, \boldsymbol{\phi}) \cdot P(r | z, \epsilon),$$

or vice versa, by correlating these two variables:

$$P(u, i, r | \Theta, \boldsymbol{\pi}) = \sum_z P(z | \boldsymbol{\pi}) \cdot P(u | z, \boldsymbol{\theta}) \cdot P(i | z, \boldsymbol{\phi}) \cdot P(r | i, z, \epsilon).$$

The corresponding graphical models are shown in figures 2.4(b) and 2.4(c).

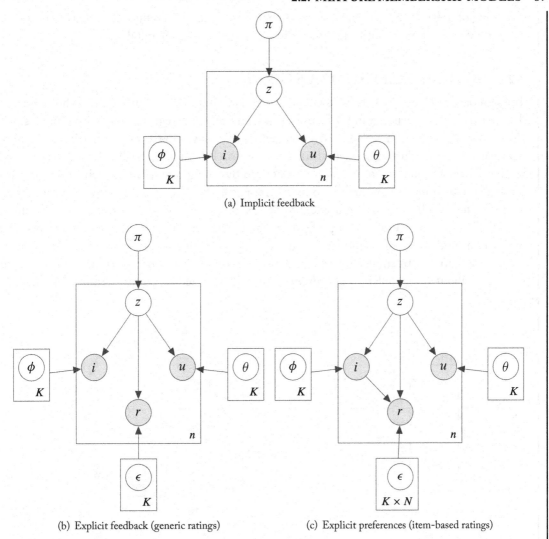

(a) Implicit feedback

(b) Explicit feedback (generic ratings)   (c) Explicit preferences (item-based ratings)

Figure 2.4: Aspect model.

Unlike the clustering model shown in the previous section, where the joint probability for a set of ratings by an individual user is modeled directly, the aspect model concentrates on the joint probability $P(u, i, r)$ separately for each rated item. As a result, different latent factors can be exploited to explain the triplet $\langle u, i, r \rangle$, whereas, the previous mixture models, all the observations for a given user are generated by the same chosen latent factor. However, the aspect model only introduces a single set of latent variables for items, users, and ratings. This essentially encodes the clustering of users, items, and the correlation between them. A step further in this direction is to try to provide multiple latent factors modeling each dimension separately.

2.2.2   MODEL-BASED CO-CLUSTERING

The approaches introduced so far focus on factorizing one single dimension of the preference data matrix by identifying groups of similar-minded users or categories of similar products. Co-clustering techniques aim at detecting homogenous blocks within the rating matrix, thus enforcing a simultaneous clustering on both its dimensions. This highlights the mutual relationships between users and items: similar users are detected by taking into account their ratings on similar items, which in turn are identified by the ratings assigned by similar users. Capturing both user and item similarities through latent variables provides more expressiveness and flexibility. Figure 2.5 shows a toy example of preference data co-clustered into blocks. In this example, the coclustering induces a natural ordering among rows and columns, thus devising blocks in the preference matrix with similar ratings. The discovery of such a structure is likely to induce information about the population and improve the personalized recommendations. Besides their

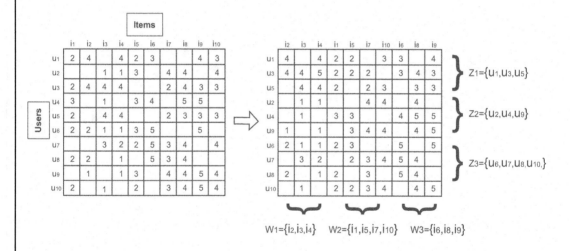

Figure 2.5: Example co-clustering for preference data.

contribution to the minimization of the prediction error, these relationships are especially important, as they can provide a faithful yet compact description of the data, which can be exploited for better decision making.

Co-clustering approaches to latent-factor modeling for two-dimensional data have been proposed in the literature, both from a matrix factorization perspective [63] and from a probabilistic perspective [70]. Here we review the original mixture model formulation, introduced in the beginning of the section, to model each preference observation as the result of a two-dimensional stochastic process governed by latent variables.

In particular, we can focus on the aspect model and consider a simple extension (called *two-sided clustering model* in [86]), which is based on the strong assumption that each person belongs to exactly one user-community and each item belong to one group of items. The rating value is independent of the user and item identities are given their respective cluster memberships. The *Flexible Mixture Model* (FMM) [98] represents a full probabilistic refinement of the above model and can be summarized as follows. Let z and w denote latent variables ranging within the values $\{1, \ldots, K\}$ and $\{1, \ldots, H\}$, representing latent factors respectively for users and items. The joint likelihood of an observation $\langle u, i, r \rangle$ can be formulated as

$$P(u, i, r | \Theta) = \sum_z \sum_w P(z|\boldsymbol{\pi}) \cdot P(w|\boldsymbol{\psi}) \cdot P(u|z, \boldsymbol{\theta}) \cdot P(i|w, \boldsymbol{\phi}) \cdot P(r|z, w, \epsilon).$$

The corresponding graphical model is shown in Figure 2.6(a). Here, the components $\boldsymbol{\pi}$ and $\boldsymbol{\psi}$ represent the multinomial probabilities relative to the latent variables z and w. Also, $\boldsymbol{\theta}_z$ and $\boldsymbol{\phi}_w$ encode the probabilities of observing users and items, respectively, given the respective latent variables, and finally, $\epsilon_{z.w}$ encodes the probability of observing ratings given z and w.

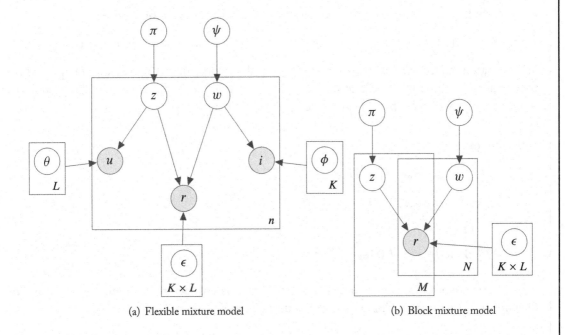

(a) Flexible mixture model   (b) Block mixture model

Figure 2.6: Graphical models for co-clustering.

FMM is a specialization of a more general formulation

$$P(u, i, r|\Theta) = \sum_z \sum_w P(r, u, i|z, w, \Theta) \cdot P(z, w|\Theta),$$

where the conditional independence assumptions $P(r, u, i|z, w, \Theta) = P(r|z, w, \Theta) \cdot P(u|z, \Theta) \cdot P(i|w, \Theta)$ and $P(z, w|\Theta) = P(z|\Theta) \cdot P(w|\Theta)$ are exploited to simplify the expression and to devise an EM strategy for learning the parameters.

This general formulation can also be adapted to forced prediction:

$$P(r|u, i, \Theta) = \sum_z \sum_w P(r|z, w, \Theta) \cdot P(z, w|u, i, \Theta). \qquad (2.8)$$

Within this formula, the term $P(z, w|u, i, \Theta)$ represents the posterior probability of the latent factors, given user u and item i. Equation 2.8 resembles the mixture equation of the single dimensional case, and an EM inference strategy can be devised as well. We adapt the results in [69, 70] to illustrate this. We assume further that preferences are expressed in a continuous domain, i.e., a preference value is the result of a Gaussian stochastic process. The adaptation to the Bernoulli or multinomial case is straightforward and we discuss it next.

A *block mixture model* (BMM) can be defined by two latent matrices $(\mathbf{Z}, \mathbf{W})$ relative to preference data, having the following characterizations:

- $\mathbf{Z} \in \{0, 1\}^{M \times K}$ and $z_{u,k} = 1$ if u belongs to the cluster k, zero otherwise;

- $\mathbf{W} \in \{0, 1\}^{N \times L}$ and $w_{i,l} = 1$ if the item i belongs to the cluster l, zero otherwise.

Given a rating matrix $\mathbf{R}$, the latent variables $z_{u,k}$ and $w_{i,l}$ determine the block (k, l) upon which to interpret the rating r_i^u associated with the pair. The underlying generative model is graphically represented in Figure 2.6(b) and can be described as follows.

1. For each u generate $z_u \sim Disc(\boldsymbol{\pi})$.

2. For each i generate $w_i \sim Disc(\boldsymbol{\psi})$.

3. For each pair $\langle u, i \rangle$:

   - Detect k and l such that $z_{u,k} = 1$ and $w_{i,l} = 1$;
   - Generate $r \sim P(r|\epsilon_{k,l})$.

Notice that, in the above model, we assume that latent factors for items and users are independent. The corresponding data likelihood can be modeled as:

$$P(\mathcal{X}, \mathbf{Z}, \mathbf{W}|\Theta) = \prod_{u \in \mathcal{U}} \prod_k \pi_k^{z_{u,k}} \prod_{i \in \mathcal{I}} \prod_l \psi_l^{w_{i,l}} \prod_{\langle u,i,r \rangle} \prod_{k,l} P(r|\epsilon_{k,l})^{z_{u,k} \cdot w_{i,l}}.$$

The estimation of the optimal parameters through a standard EM procedure is difficult here, because it requires inferring $P(\mathbf{Z}, \mathbf{W}|\mathcal{X}, \Theta)$ and, in particular, the term $P(z_{u,k} w_{i,l} = 1|u, i, r, \Theta)$ relative to an observation $\langle u, i, r \rangle$ in $\mathcal{X}$. A solution is provided in [69], in terms of *variational approximation*. The key idea is to approximate an intractable distribution $P(\Theta|\mathcal{X})$ with a simple, but tractable, distribution $Q(\Theta|\mathcal{X})$ and then to minimize their Kullback-Leibler divergence $KL(Q, P)$ (for a general introduction to variational inference see Appendix A.2). In this case, we devise a generalized EM procedure by introducing a mathematically tractable approximation $q(\mathbf{Z}, \mathbf{W})$ to the posterior $P(\mathbf{Z}, \mathbf{W}|\mathcal{X}, \Theta)$. By exploiting the factorization

$$q(\mathbf{Z}, \mathbf{W}) = P(\mathbf{Z}|\mathcal{X}, \Theta) P(\mathbf{W}|\mathcal{X}, \Theta),$$

and using the notation $c_{u,k} \triangleq P(z_{u,k} = 1|u, \Theta')$, $d_{i,l} \triangleq P(w_{i,l} = 1|i, \Theta')$, we can redefine the complete-data expectation log-likelihood $\mathcal{Q}(\Theta, \Theta')$ as follows:

$$\mathcal{Q}(\Theta, \Theta') = \sum_{k=1}^{K} \sum_{u} c_{u,k} \log \pi_k + \sum_{l=1}^{L} \sum_{i} d_{i,l} \log \psi_l + \sum_{\langle u,i,r \rangle} \sum_k \sum_l \left\{ c_{u,k} \cdot d_{i,l} \log P(r|\epsilon_{k,l}) \right\}.$$

The terms $c_{u,k}$ and $d_{i,l}$ represent the standard responsibilities computed on a single dimension (users for $c_{u,k}$ and columns for $d_{i,l}$).[5] For the M step, instead of directly optimizing $\mathcal{Q}(\Theta, \Theta')$, it is convenient to look for a generic improvement $\hat{\Theta}$ such that $\mathcal{Q}(\hat{\Theta}, \Theta') \geq \mathcal{Q}(\Theta, \Theta')$. This can be done by decomposing $\mathcal{Q}(\Theta, \Theta')$, by alternatively assuming that the pairs $(\mathbf{c}, \pi)$ (resp. $(\mathbf{d}, \psi)$) are fixed, and by optimizing the remaining parameters through a standard M step.

The core of the approach is the possibility of refactoring $\mathcal{Q}(\Theta, \Theta')$ by exploiting sufficient statistics for the computation of $P(r|\epsilon_{k,l})$. As stated above, we assume a Gaussian distribution for the ratings, so that $\epsilon_{k,l} \triangleq \{\mu_k^l, \sigma_k^l\}$, and $P(r|\epsilon_{k,l}) \triangleq \mathcal{N}(r; \mu_k^l, \sigma_k^l)$. We can notice [18] that the joint probability of a generic normal population $\{x_1, \ldots, x_n\}$ can be factored as:

$$\prod_{i=1}^{n} \mathcal{N}(x_i; \mu, \sigma) = h(x_1, \ldots, x_n) \cdot \varphi(u_0, u_1, u_2; \mu, \sigma),$$

where

$$h(x_1, \ldots, x_n) = (2\pi)^{-n/2}, \qquad \varphi(u_0, u_1, u_2; \mu, \sigma) = \sigma^{-u_0} \cdot \exp\left(\frac{2 \cdot u_1 \cdot \mu - u_2 - u_0 \cdot \mu^2}{2 \cdot \sigma^2} \right)$$

and $u_0 = n$, $u_1 = \sum_i x_i$ and $u_2 = \sum_i x_i^2$ are the sufficient statistics. Based on the above observation, we can decompose $\mathcal{Q}(\Theta, \Theta')$ as

$$\mathcal{Q}(\Theta, \Theta') = \mathcal{Q}(\Theta, \Theta'|\mathbf{d}) + \sum_{i \in \mathcal{I}} \sum_{l=1}^{L} d_{i,l} \log \psi_l - \sum_{u \in \mathcal{U}} \sum_{i \in \mathcal{I}(u)} \frac{d_{i,l}}{2 \log(2\pi)},$$

[5]These terms can be determined by solving the corresponding equations adapted from Equation A.9. As described in Appendix A.2, this variational principle is equivalent to minimizing the KL-divergence between the approximation and the true posterior distributions.

where

$$\mathcal{Q}(\Theta, \Theta'|\mathbf{d}) = \sum_{u=1}^{M} \sum_{k=1}^{K} c_{uk} \left\{ \log(\pi_k) + \sum_{l=1}^{L} \log\left(\varphi(u_0^{(u,l)}, u_1^{(u,l)}, u_2^{(u,l)}; \mu_k^l, \sigma_k^l)\right) \right\}$$

and

$$u_0^{(u,l)} = \sum_{i \in \mathcal{I}(u)} d_{il}; \quad u_1^{(u,l)} = \sum_{i \in \mathcal{I}(u)} d_{il} r_i^u; \quad u_2^{(u,l)} = \sum_{i \in \mathcal{I}(u)} d_{il} (r_i^u)^2.$$

Analogously,

$$\mathcal{Q}(\Theta, \Theta') \;=\; \mathcal{Q}(\Theta, \Theta'|\mathbf{c}) + \sum_{u \in \mathcal{U}} \sum_{k=1}^{K} c_{u,k} \cdot \log \pi_k - \sum_{i \in \mathcal{I}} \sum_{u \in \mathcal{U}(i)} \frac{c_{u,k}}{2 \log(2\pi)},$$

where

$$\mathcal{Q}(\Theta, \Theta'|\mathbf{c}) = \sum_{i=1}^{N} \sum_{l=1}^{L} d_{i,l} \cdot \left\{ \log(\psi_l) + \sum_{k=1}^{K} \log\left(\varphi(u_0^{(i,k)}, u_1^{(i,k)}, u_2^{(i,k)}; \mu_k^l, \sigma_k^l)\right) \right\}$$

and

$$u_0^{(i,k)} = \sum_{u \in \mathcal{I}(u)} c_{u,k}; \quad u_1^{(i,k)} = \sum_{u \in \mathcal{I}(u)} c_{u,k} \cdot r_i^u; \quad u_2^{(i,k)} = \sum_{u \in \mathcal{I}(u)} c_{u,k} \cdot (r_i^u)^2.$$

Thus, by assuming ψ and $\mathbf{d}$ constant, and optimizing $\mathcal{Q}(\Theta, \Theta')$ with respect to the remaining parameters, still guarantee a new value $\hat{\Theta}$ such that $\mathcal{Q}(\hat{\Theta}, \Theta') \geq \mathcal{Q}(\Theta', \Theta')$. Dually, assume ϕ and $\mathbf{c}$ are fixed and obtain an improvement on the expectation-likelihood by updating the remaining parameters.

The overall procedure can be summarized as follows.

- Start with initial values of $\mathbf{c}^{(0)}$, $\mathbf{d}^{(0)}$ and $\Theta^{(0)}$.

- Iterate until convergence:

  - Compute $(\mathbf{c}^{(t+1)}, \pi^{(t+1)}, \Theta^{(t')})$ by optimizing $\mathcal{Q}(\Theta, \Theta^{(t)}|\mathbf{d}^{(t)})$;
  - Compute $(\mathbf{d}^{(t+1)}, \psi^{(t+1)}, \Theta^{(t+1)})$ by optimizing $\mathcal{Q}(\Theta, \Theta^{(t')}|\mathbf{c}^{(t+1)})$.

The optimization steps are straightforward, and they are omitted here. The interested reader can find the details in [18].

Although the above model only focuses on Gaussian modeling, it can be generalized to other probability distributions that can be expressed in terms of sufficient statistics. In particular, Agarwal et al. [4] adapt the above model to a framework of generalized linear models, which include commonly used regression models such as linear, logistic, and Poisson regression as special cases.

2.3   PROBABILISTIC LATENT SEMANTIC MODELS

Probabilistic topic models [32, 34, 188] include a suite of techniques widely used in text analysis: they provide a low-dimensional representation of the given data, which allows the discovering of *semantic relationships*. Given a collection, said *corpus* of macro entities to be modeled (textual documents or users' purchase logs), topic models are based upon the idea that each entity exhibits a mixtures of topics, where a topic is a probability distribution over a finite set of elements (e.g., tokens or items). The underlying generative process assumes that, to generate an entity, we first sample a distribution over topics. Each single data observation, generally referred to as token, can be generated by sampling a topic from the specific distribution and then emitting a token by drawing upon the topic distribution. Bayesian inference techniques, discussed next, are used to infer a set of topics from a collection of observations, by "reversing" the generative model and hence determining the distributions that are more likely to generate the observed data.

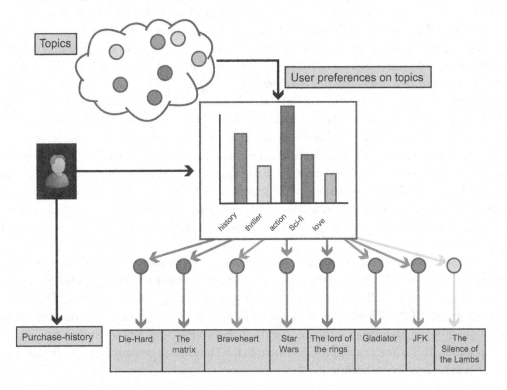

Figure 2.7: Latent class model for CF—generative process.

The adaptation of topic modeling to CF is quite natural. For example, Figure 2.7 illustrates the interpretation of CF data in terms of topic modeling. Here, colors represent topics and the histogram represents their distribution relative to a specific user. Although no prior conceptual meaning can be associated to topics, in the case of CF data it is natural to assume that they correspond to users' interests. A proper definition of topics might be obtained by considering them as "abstract preference pattern:" users, or items, participate in each preference pattern with a certain degree of involvement, and these membership weights project each user/item into the latent factor space. We assume that there are a fixed number of topics, and each user is characterized by her own preference on genres. For example, in Figure 2.7, the considered user shows a particular interest in action and historical movies, while her interest in romance is low. Each genre specifies the probability of observing each single item. Movies like "Independence Day" and "Die hard" will have a higher probability of being observed given the "action" topic, than in the context of "romance." Given a user and her preferences on topics (which defines preferences on movies), the corresponding purchasing history can be generated by choosing a topic and then drawing an item from the corresponding distribution over items. In the example, the first topic to be chosen is "action," which generates the movie "Die Hard;" the process of topic and item selection is iteratively repeated to generate the complete purchase history of the current user.

We shall explore the exploitation of topic models to preference data in two respects. A first approach draws from the aspect model proposed before and focuses on an alternate, probabilistic interpretation of Latent Semantic Indexing (LSI) in terms of topic modeling. This approach is named *Probabilistic Latent Semantic Analysis/Indexing (pLSA/pLSI)*. The intuition behind LSI is to find the latent structure of "topics" or "concepts" in a text corpus, which captures the meaning of the text that is imagined to be obscured by "choice" noise. Applied to preference data, these concepts can suitably model preference choice in terms of complex latent factor elements.

A generalization of the pLSA approach is the *Latent Dirichlet allocation* (LDA), introduced by Blei et al. [34] and discussed in the next chapter. LDA overcomes the over-specialization of pLSA (the topic mixture component is known only for those users u in the training set) by providing a full generative semantic. The relationship between these two approaches is strong, as it was shown that pLSA is a maximum *a posteriori* estimated LDA model under a uniform Dirichlet prior [65].

The intuition behind PLSA is to reinterpret the probability of a preference $P(r|u, i)$ in terms of a mixture components, where each component encodes a specific parametric model. An alternative view is to still model $P(r|u, i)$ as a specific parametric distribution. However, it is the parameters of this distribution that can be influenced by latent factors, so that the semantic decoupling of the preference is focused on the parameter setting rather then on the probability components. This is a second approach known in the literature as *probabilistic matrix factorization*, which will be detailed in Section 2.3.2.

2.3.1   PROBABILISTIC LATENT SEMANTIC ANALYSIS

pLSA [81, 83] specifies a co-occurrence data model in which the user u and item i are conditionally independent, given the state of the latent factor z. In contrast to standard latent semantic analysis based on SVD decomposition, pLSA defines a proper generative data model, which has several advantages. Statistical inference techniques, like likelihood maximization, can be easily used to learn the model, to minimize directly the accuracy of predictions over a test set (as we shall discuss in Chapter 4). Secondly, the probabilistic framework allows us to employ model selection techniques to control the complexity and the fitting of the model, and finally, experimental results show that pLSI achieves substantial performance gains over LSA [81].

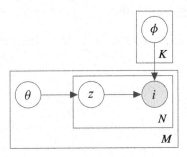

Figure 2.8: Probabilistic latent semantic analysis.

Figure 2.8 details the instantiation of pLSA to implicit preferences. Here, θ is a prior distribution of topics associated to each user. For a given user u, a multinomial distribution θ_u is defined over the latent space of topics. Furthermore, each item is chosen from a multinomial distribution ϕ_k, relative to the current topic k and ranging over all possible items.

It is important not to confuse the pLSA model with probabilistic clustering models. In these models, a single latent factor is associated with each user. By contrast, the pLSA model associates a latent variable with each observation triplet $\langle u, i, r \rangle$. Hence, different observations for the same user can be explained by different latent causes in pLSA, whereas a user clustering model assumes that all ratings involving the same user are linked to the same underlying community/cluster. This is better illustrated by the underlying generative process:

- For each user u:

  - Sample the number n_u of item selections;
  - For each selection $s \in \{1, \ldots, n_u\}$:
    1. Select a user profile $z \sim Disc(\theta_u)$;
    2. Pick an item $i \sim Disc(\phi_z)$.

pLSA shares many similarities with the aspect model discussed in the previous section. Both models associate latent factors to single observations, rather than to users. Also, the probability of observing an adoption is governed in both models by a distribution over all the possible items. However, the aspect model fucuses on free prediction and aims at directly modeling the joint probability $P(u, i)$, whereas pLSA considers the probability $P(i|u)$, which is modeled as a mixture

$$P(i|u) = \sum_k \theta_{k,u} \phi_{k,i}.$$

Given a set $\mathcal{X}$ of observations, the likelihood can be specified by assuming a latent matrix $\mathbf{Z}$, where e generic entry $z_m \in \{1, \dots, K\}$ is relative to the observation $m = \langle u, i \rangle$ and represents the latent factor underlying such an observation. The joint likelihood can be expressed hence as

$$P(\mathcal{X}, \mathbf{Z}|\Theta) = P(\mathcal{X}|\mathbf{Z}, \Theta) P(\mathbf{Z}|\Theta),$$

where

$$P(\mathcal{X}|\mathbf{Z}, \Theta) = \prod_{\langle u,i \rangle} \phi_{z_{u,i},i},$$

and

$$P(\mathbf{Z}|\Theta) = \prod_{\langle u,i \rangle} \theta_{u,z_{u,i}}.$$

As usual, the likelihood can be obtained by marginalizing over all possible $\mathbf{Z}$:

$$L(\Theta; \mathcal{X}) = \sum_{\mathbf{Z}} P(\mathcal{X}, \mathbf{Z}|\Theta) = \prod_{\langle u,i \rangle} \sum_k \theta_{u,k} \cdot \phi_{k,i}. \tag{2.9}$$

Again, the standard EM framework can be exploited to find the optimal Θ.

The adaptation of the model to explicit preferences can be modeled as well, in a way similar to the AM discussed above. For example, we can adopt a multinomial distribution $\epsilon_{k,i}$ [82], such that

$$P(r|u, i) = \sum_{k=1}^K \epsilon_{i,k,r} \cdot \theta_{u,k}. \tag{2.10}$$

Alternatively, the *Gaussian PLSA (G-PLSA)* [84]) models $\epsilon_{k,i} = (\mu_{i,k}, \sigma_{i,k})$ as a Gaussian distribution. The corresponding rating probability is

$$P(r|u, i) = \sum_{k=1}^K \mathcal{N}(r; \mu_{i,k}, \sigma_{i,k}) \cdot \theta_{u,k}. \tag{2.11}$$

It is well known that statistical models involving huge numbers of parameters are prone to overfitting. pLSA represents one such model and, hence, it is not immune to this issue. Traditional model selection techniques can be employed to mitigate such a drawback, by fitting models by maximum likelihood and then determining the generalization performance of the model either analytically, by directly exploiting MAP in place of maximum likelihood, or via empirical evaluation using hold-out data or cross-validation. A more rigorous framework is offered by Bayesian learning and will be discussed in the next chapter.

2.3.2   PROBABILISTIC MATRIX FACTORIZATION

We have seen that the basic intuition behind pLSA is to reinterpret the probability of a preference $P(r|u,i)$ in terms of a mixture of components $P(r|u,i) = \sum_z P(r|i,z)P(z|u)$. Here, z represents a latent factor and the mixture reinterprets the matrix approximation discussed in Section 1.4.2, in probabilistic terms.

An alternative approach consists instead of "moving" the matrix approximation to the parameter set. That is, we can still model $P(r|u,i)$ as a specific parametric distribution $P(r|u,i,\Theta_{u,i})$. However, the parameter set $\Theta_{u,i}$, in this case, can depend on some latent factors, which hence influence the parametric distribution directly. Consider again the two alternatives. For the continuous case, we can assume a Gaussian distribution,

$$P(r|u,i) = \mathcal{N}(r; \mu_{u,i}, \sigma),$$

where, for simplicity, we assume a single observation-specific parameter $\mu_{u,i}$. For the discrete multinomial case, we can assume the multinomial probability $\{p_{u,i,v}\}_{v \in \mathcal{V}}$ such that

$$P(r|u,i) = p_{u,i,r}.$$

Let us focus on the continuos case for the moment. We can make explicit a dependency of the parameter $\mu_{u,i}$ from two continuous latent factors $\mathbf{P}_u, \mathbf{Q}_i \in \mathbb{R}^K$, so that $\mu_{u,i} \triangleq \mathbf{P}_u^T \mathbf{Q}_i$. This approach is called *Probabilistic Matrix Factorization (PMF)* and was introduced in [169]. Given the latent user and item feature vectors $\mathbf{P}_u, \mathbf{Q}_i$, the preference value is generated by assuming a Gaussian distribution over rating values conditioned on the interactions between the user and the considered item in the latent space, as shown below:

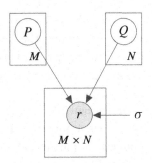

To summarize, the continuous case $P(r|u,i)$ is modeled as a Gaussian distribution, with mean $\mathbf{P}_u^T \mathbf{Q}_i$ and fixed variance σ:

$$P(r|u,i,\mathbf{P},\mathbf{Q}) = \mathcal{N}(r; \mathbf{P}_u^T \mathbf{Q}_i, \sigma). \tag{2.12}$$

Similarly, in the discrete case we can assume that the parameter $p_{u,i,r}$ depends on some latent factors $\mathbf{P}_{r,u}, \mathbf{Q}_{r,i} \in \mathbb{R}^K$, i.e., $p_{u,i,r} \propto \mathbf{P}_{r,u}^T \mathbf{Q}_{r,i}$. This yields to PMF reformulation of the probability of a rating as

$$P(r|u, i, \mathbf{P}, \mathbf{Q}) = \frac{\exp\left\{P_{r,u}^T Q_{r,i}\right\}}{\sum_{v \in \mathcal{V}} \exp\left\{P_{v,u}^T Q_{v,i}\right\}}. \tag{2.13}$$

There is an interesting parallel between PMF and the predictive models shown in Section 2.1. If we consider the equations 2.2 and 2.3 under a more general perspective, we can reformulate them as depending on a functional $\Psi(u, i, r; \Theta)$, which can be embedded either into the Gaussian or the logistic probability. For example, we can reformulate Equation 2.2 as

$$P(r|u, i, \Theta) = \frac{1}{\sqrt{2\pi}\sigma} \exp\left\{-\frac{1}{2\sigma^2}(r - \Psi(u, i, r; \Theta))^2\right\}. \tag{2.14}$$

Under this perspective, PMF is just a different instantiation of the functional $\Psi(u, i, r; \Theta)$. Indeed, in Equation 2.2 we instantiated it as $\Psi(u, i, r; \Theta) \triangleq \sum_j \beta_j f_j(u, i, r)$, but, if we replace this instantiation with $\Psi(u, i, r; \Theta) \triangleq \mathbf{P}_u^T \mathbf{Q}_i$, we can then observe that equations 2.14 and 2.12 are equivalent. The same parallel holds for the discrete case.

In the PMF formulation, the likelihood can be expressed as

$$P(\mathcal{X}|\mathbf{P}, \mathbf{Q}, \sigma) = \prod_{\langle u, i, r \rangle} P(r|u, i, \mathbf{P}, \mathbf{Q}).$$

The estimation of the $\mathbf{P}$ and $\mathbf{Q}$ parameters via MLE can be accomplished again by resorting to iterative methods, as discussed in Section 2.1. If we consider the log-likelihood of the Gaussian case,

$$\log P(\mathcal{X}|\mathbf{P}, \mathbf{Q}, \sigma) = -\frac{1}{2\sigma} \sum_{\langle u, i, r \rangle} (r - \mathbf{P}_u^T \mathbf{Q}_i)^2 - n \log \sigma - \frac{n}{2} \log 2\pi,$$

we can observe the equivalence between optimizing the latter and Equation 1.22. That is, the Gaussian PMF provides a probabilistic interpretation of the matrix factorization discussed in Chapter 1, which is based on minimizing the sum-of-squared-errors objective function with quadratic regularization terms.

Different extensions of the basic framework have been proposed: a constrained version [169] of the PMF model is based on the assumption that users who have rated similar sets of items are likely to exhibit similar preferences. Furthermore, Bayesian generalizations [168] and extensions with side information [1, 180, 209] are characterized by a higher prediction accuracy. We shall analyze them in the following chapter. Also, in Chapter 6 we shall review and further extend the models proposed here to include side information.

2.4   SUMMARY

Within this chapter we investigated the different approaches to model preference data in a probabilistic setting. We started our treatment by distinguishing between forced prediction (aimed at modeling the probability $P(r|u,i)$ and free prediction (aimed at modeling either $P(r,i|u)$ or $P(u,i,r)$). The difference between these modeling choices resembles the difference between *discriminative* and *generative* modeling. In general, discriminative modeling aims at focusing on prediction accuracy. By converse, generative modeling is more general: a free prediction model naturally induces a forced prediction model as well, as $P(r|u,i) = P(r,i|u)/\sum_{i\in\mathcal{I}} Pr(r,i,u)$. Thus, the informative content of a generative model is richer, although at the price of a higher computational learning cost and a lower accuracy in predicting the rating. Thus, a first dimension upon which to consider the modeling is either free or forced prediction, and different choices produce different models.

Another prominent role is played by latent factor modeling. The underlying idea of latent factor modeling is to reinterpret the data into a low-dimensional representation, which allows the discovering of *semantic relationships*. This enables the detection of similarities among users and/or items and, in general, allows the identification of a fixed number of more general preference patterns. Again, latent factors can be exploited in two respects. Either in a "discrete" fashion, by assuming that a latent factor represents a discrete state and a single preference can be observed in any of these states; or, alternatively, in a continuous domain, as the result of a process that directly affects the parameters of the underlying data distribution.

The following table summarizes the approaches presented in this chapter under these two dimensions.

Table 2.1: Summary

		MMM	UCM	AM	BMM	FMM	pLSA	G-PLSA	PMF
PREDICTION MODE	Free		✓	✓		✓			
	Forced	✓			✓		✓	✓	✓
FACTORIZATION DIMENSIONS	User-based	✓	✓	✓	✓		✓	✓	
	Item-based				✓				
	Joint					✓			✓
FACTORIZATION MODEL	Discrete	✓	✓	✓	✓	✓	✓	✓	
	Continuous								✓
FACTORIZATION SCOPE	Hard	✓	✓		✓				
	Soft			✓			✓	✓	✓
PREFERENCE MODELING	Implicit feedback			✓			✓		
	Explicit (Multinomial)	✓	✓	✓		✓			
	Explicit (Gaussian)	✓	✓	✓	✓			✓	✓

Within this table, we can see that several aspects characterize each dimension: for example, latent topics can group users, items, or both dimensions. Also, the modeling of the explicit rating can be either continuous or discrete. Again, some models naturally fit in either the first, the second, or both of the modeling alternatives. As we will see in Chapter 4, probabilistic models that are based on the estimate of Gaussian distributions over rating values outperform significantly, in a rating prediction scenario, approaches in which the generative distribution is multinomial.

A final dimension to consider is the scope of the latent factor modeling. We have seen that simple models associate a latent variable to a user; as a consequence, all preferences associated to that user can be explained in terms of that latent variable. This is what we name a *hard* choice in the table. However, a different alternative is to consider the influence of latent factors at the level of a single preference, and all the components of an observation can be thought of as the result of a stochastic process influenced by the latent factors. This is the *soft* scope.

The maximum likelihood estimation procedures corresponding to these models are easy to implement. However, the two main drawbacks of the approaches described in this chapter are the tendency to overfit and the sensitivity to the initial conditions in the learning procedure. For the latter, two common practices are to perform several runs corresponding to different random initial configurations of parameters, or to determine a "good" initial setting of parameters. For the former, it is common to observe that the fitting procedures, which notably are guaranteed to perform iterative updates on the parameters of the considered model that increase the log likelihood on training data, tend to deteriorate the performances on unseen data for increasing numbers of iterations. Early stopping criteria, as well as model selection strategies, can be employed to mitigate such phenomena from a practical point of view. A more rigorous framework is instead offered by a Bayesian treatment and will be discussed in the next chapter.

CHAPTER 3

Bayesian Modeling

The methods introduced in the previous chapter rely on maximum likelihood estimation and make no assumption on the likelihood of Θ. That is, all Θ are equally probable and, thus, the optimal parameter set is uniquely identified by the observed data. An alternative approach assumes that we can incorporate prior knowledge about the domain of Θ. Prior knowledge can be combined with observed data to determine the final optimal parameter set $\hat{\Theta}$. Rather than optimizing the likelihood, we concentrate on the probability $P(\Theta|\mathcal{X})$ and seek the set of parameters that maximizes it. By exploiting Bayes' theorem, the likelihood can be expressed as

$$P(\Theta|\mathcal{X}) = \frac{P(\mathcal{X}|\Theta)P(\Theta)}{\int P(\mathcal{X}|\Theta)P(\Theta)\mathrm{d}\Theta}. \tag{3.1}$$

Since the optimal value only depends on the numerator, we can reformulate the solution of the estimation problem as a *maximum a posteriori probability* (MAP) estimate:

$$\hat{\Theta}_{MAP} = \underset{\Theta}{\operatorname{argmax}} \, P(\Theta|\mathcal{X}) = \underset{\Theta}{\operatorname{argmax}} \, P(\mathcal{X}|\Theta)P(\Theta),$$

the MAP estimate allows us to inject into the estimation calculation our prior beliefs regarding the parameter values in Θ. The integration of $P(\Theta)$ realizes a Bayesian approach to data modeling, where we can state beliefs on how "realistic" are some parameters with respect to others. Bayesian modeling is particularly useful when observations $\mathcal{X}$ do not provide sufficient evidence: in these cases, the estimate is driven by the prior component. Finally, given $\hat{\Theta}_{MAP}$, we can solve the prediction problem as:

$$P(\tilde{x}|\mathcal{X}) \approx P(\tilde{x}|\hat{\Theta}_{MAP}).$$

Let us consider again the simple example discussed at the beginning of Chapter 2. Recall that observations $\mathcal{X}$ represent observations that can be obtained from the preference matrix $\mathbf{R}$, and a Bernoulli distribution governing the presence/absence of preference, parameterized by $\Theta \triangleq \mu$. Within MAP estimation, we can instantiate $P(\Theta)$ by using the Beta distribution $P(\Theta|a,b) = B(a,b)^{-1}\mu^a(1-\mu)^b$, which is the *conjugate prior* of the Bernoulli distribution [31, 62, Chapter 2]. A prior $P(\Theta)$ and a posterior distribution $P(\Theta|\mathcal{X})$ are conjugate distributions if they belong to the same family; in this case the prior is said to be a conjugate prior for the likelihood function $P(\mathcal{X}|\Theta)$. The advantage in using conjugate priors relies on algebraic convenience, as the posterior can be expressed in closed form. For instance, in our example, where the likelihood is modeled as a Bernullian distribution and the prior as Beta, the posterior easily can be rewritten as

$$P(\Theta|\mathcal{X}) = B(n + a, M \times N - n + b)^{-1} \mu^{n+a}(1 - \mu)^{M \times N - n + b},$$

which yields

$$\hat{\mu}_{MAP} = \frac{n + a}{M \times N + a + b}.$$

Here, $B(a, b)$ represents the Beta function. The Bayesian inference in the case of explicit preference can be obtained in a similar way, by properly choosing the prior $P(\Theta)$: e.g., by adopting a normal prior in case of Gaussian likelihood, or a Dirichlet prior when the generative distribution is multinomial [58].

There is an immediate relationship between MAP and ML estimations, stated by Bayes's theorem:

$$posterior \propto likelihood \times prior.$$

In both cases, the likelihood component plays a crucial role but the way in which it is used is fundamentally different in the two approaches. In a ML setting, the parameter set is considered fixed and the variability in the estimation can be evaluated by considering the distribution of all possible observations, of which $\mathcal{X}$ represents a (possibly biased) sample. By contrast, from a Bayesian perspective, the observations $\mathcal{X}$ are fixed, and the uncertainty in the parameters is expressed through a probability distribution over the parameter set Θ. The advantage is that, by incorporating the prior, we can mitigate the effects of bias within $\mathcal{X}$, typically due to sparsity issues associated with the sample under consideration.

In this chapter, we review the modeling paradigms proposed in the previous chapter under a Bayesian perspective. A first advantage of a Bayesian modeling is that it allows a smoothing of the estimation procedure. For example, it allows us to identify the effective number of latent factors, as well as deal with the sparsity of the preference matrix. In a sense, we are regularizing the parameter set by constraining the optimal values in a specific range. Put in other terms, including the prior in the estimation allows us a better control of the overfitting issues, thus resulting in models with better generalization capabilities.

However, there's more in Bayesian modeling than just regularization. Bayesian inference extends the MAP approach by allowing a distribution over the parameter set Θ, instead of making a direct estimate. Thus, it is more informative: it still encodes the maximum (a posteriori) value of the data-generated parameters, but it also incorporates expectation as another parameter estimate. This enables the possibility to approach the prediction problem by more reliable inference approximations. A MAP estimator is popular in Bayesian analysis, in part, because it is often computationally less demanding than computing the posterior mean or the median. The reason is simple; to find the maximum, the posterior need not to be fully specified. However, the posterior is the ultimate experimental summary in a Bayesian perspective.

3.1   BAYESIAN REGULARIZATION AND MODEL SELECTION

A problem with maximum likelihood is the susceptibility to overfitting, especially when the data at hand is sparse. Overfitting can be roughly defined as an exceptionally good capability of a model of describing the observations in $\mathcal{X}$ and, by the converse, poor predictive capability on unseen samples $\tilde{\mathbf{x}}$. The complexity of Θ could enforce overfitting. Complex parameter sets exhibit larger degrees of freedom and, especially when the observations are sparse, this may lead to poor estimations. For a given set of observations $\mathcal{X}$, we can consider the "true" probability $P(\mathcal{X})$ independent of the parameter, i.e.

$$P(\mathcal{X}) = \int P(\mathcal{X}|\Theta) P(\Theta) \, d\Theta.$$

The "optimal" parameter set is the one that best approximates the integral on the right-hand side. ML estimation provides an approximation by only looking at the likelihood $P(\mathcal{X}|\Theta)$ and it ignores the $P(\Theta)$ component. By the converse, MAP estimation does a better job by also considering the smoothing effect of the prior. This smoothing effect also plays the role of regularizing the choice of the optimal parameters, by binding them to reliable values.

Consider again the log-linear model discussed in Section 2.1. When the feature set $\mathcal{F}$ contains an extremely huge number of features, then the number of parameters $\beta_{r,k}$ can be extremely high. Combined with the sparsity of the feature matrix, this effect is likely to produce situations where the learned parameters are more sensitive to noise in the data. This may result in a poor predictive capability of the resulting model. Prior probabilities can help in this setting, since they allow us to encode a prior belief about the suitable values, and hence constrain the search space in the parameter estimation phase. In particular, a huge feature space $\mathcal{F}$ can be characterized by redundant and irrelevant features, and the learning phase should instead focus on the selection of the smallest subset of such features that effectively discriminate in the prediction process. This can be achieved by modeling the fact that each weight $\beta_{r,k}$ should be close to zero, unless there is strong evidence in the observation about its discriminative abilities. We can encode such an intuition within a Gaussian prior with 0 mean and fixed precision,

$$P(\beta_{r,k}) \sim \mathcal{N}(0, \sigma),$$

thus obtaining the log posterior

$$\log P(\Theta|\mathcal{X}) \propto LL(\Theta|\mathcal{X}) - \frac{1}{2\sigma^2} \sum_{r \in \mathcal{V}} \sum_{k=1}^{q} \beta_{r,k}^2.$$

Here, the term $-1/2\sigma^2 \sum_{r,k} \beta_{r,k}^2$ acts as a regularization term which penalizes the likelihood when the components $\beta_{r,k}$ approach extreme values. Hence, optimizing this objective function yields smooth values for the parameters, which are less prone to overfitting.

A similar regularization can be devised for the PMF models discussed in Section 2.3.2. Consider again the likelihood

$$P(\mathcal{X}|\mathbf{P}, \mathbf{Q}, \sigma) = \prod_{\langle u,i,r \rangle} P(r|u, i, \mathbf{P}, \mathbf{Q}).$$

Optimizing the above likelihood with respect to the matrices $\mathbf{P}$ and $\mathbf{Q}$ can readily lead to overfitting. To avoid this, we can place zero-mean spherical Gaussian priors on user and item feature vectors

$$P(\mathbf{P}|\sigma_U) = \prod_{u=1}^{M} \mathcal{N}(\mathbf{P}_u; 0, \sigma_U^2 \mathbf{I}), \qquad P(\mathbf{Q}|\sigma_I) = \prod_{i=1}^{N} \mathcal{N}(\mathbf{Q}_i; 0, \sigma_I^2 \mathbf{I}).$$

Combining the likelihood with the priors and taking the logarithm yields the log posterior:

$$\log P(\mathbf{P}, \mathbf{Q}|\mathcal{X}, \sigma, \sigma_U, \sigma_I) = -\frac{1}{2\sigma} \sum_{\langle u,i,r \rangle} (r - \mathbf{P}_u^T \mathbf{Q}_i)^2$$
$$-\frac{1}{2\sigma_U} \sum_u \mathbf{P}_u^T \mathbf{P}_u - \frac{1}{2\sigma_I} \sum_u \mathbf{Q}_i^T \mathbf{Q}_i$$
$$- n \log \sigma^2 - MK \log \sigma_U^2 - NK \log \sigma_I^2 + C,$$

where C is a constant that does not depend on the parameters. By denoting $\lambda_U = \sigma^2/\sigma_U^2$ and $\lambda_I = \sigma^2/\sigma_I^2$, we can rewrite the negative log-posterior as:

$$E(\mathbf{P}, \mathbf{Q}) = \frac{1}{2} \sum_{\langle u,i,r \rangle} (r - \mathbf{P}_u^T \mathbf{Q}_i)^2 + \frac{1}{2}\lambda_U \sum_u \mathbf{P}_u^T \mathbf{P}_u + \frac{1}{2}\lambda_I \sum_u \mathbf{Q}_i^T \mathbf{Q}_i,$$

thus obtaining an error function that resembles the one corresponding to the standard matrix factorization approach introduced in Equation 1.24 of Chapter 1. An optimization based on the gradient-descent of the above error function yields exactly the update Equations 1.25.

As part of the Bayesian framework, prior distributions describe the prior beliefs about the properties of the function being modeled. These beliefs are updated after taking into account observational data by means of a likelihood function that connects the prior beliefs to the observations. Under this respect, priors can be accommodated to model different situations according to the known facts about data. For example, in the PMF approach, more complex priors can be devised, by exploiting Gaussian distribution for more general means μ_U, μ_I and covariance matrices Σ_U, Σ_I. This allows us to model more complex situations, like, e.g., the ones described in [197], where regularization can be adapted to specific user/item cases. We shall discuss this extension in detail in Section 3.4.

Bayesian regularization can also be exploited for model selection purposes. The Bayesian view of model selection involves the use of probabilities to represent uncertainty in the choice of the model. Assume we can devise two alternative models Θ^1 and Θ^2, corresponding to a stochastic process for which the observed data is given by $\mathcal{X}$. The uncertainty in these models is given by the prior probability $P(\Theta^i)$, according to which the posterior can be written as

$$P(\Theta^i | \mathcal{X}) \propto P(\mathcal{X} | \Theta^i) P(\Theta^i).$$

The choice between Θ^1 and Θ^2 can be made by comparing their posteriors. The model of choice is the one with the larger value of the posterior.

A specific example where Bayesian modeling supports model selection is given by the choice of the number of components in latent factor models. Assume that Θ can be parametrized by K factors, i.e., there are enumerable possible parameters $\{\Theta^{(K)}\}_{K \geq 0}$ and $\Theta^{(K)} = \{\Theta_1^{(K)}, \dots, \Theta_K^{(K)}\}$. We also assume that $P(\Theta^{(K)}) = \prod_{k=1}^K P(\Theta_k^{(K)})$. This yields the joint likelihood

$$\log P(\mathcal{X}, \Theta^{(K)}) = \log P(\mathcal{X} | \Theta^{(K)}) + \sum_{k=1}^K \log P(\Theta_k^{(K)}).$$

The above situation sheds some light on the role of the number of components in a Bayesian perspective: the posterior depends on the likelihood and on the number of parameters. These dependencies can be made explicit by exploiting the improper prior[1] $P(\Theta_k^{(K)}) \propto w^{-\alpha}$ with w constant. This prior in the above equation can be rewritten by adding a penalization factor that depends on the number of components and counterbalances the likelihood: as

$$\log P(\mathcal{X}, \Theta^{(K)}) \propto \log P(\mathcal{X} | \Theta^{(K)}) - \alpha K \log w.$$

Thus, the best model can be devised as the one that provides the best compromise between the number of components K and the likelihood of the data.

The selection of the optimal number K^* of components can be plugged even within the EM framework itself. Consider the simple mixture model, where each user is associated to a latent variable, and the observation associated to each user then can be explained in terms of the associated model component. We rewrite the complete-likelihood as

$$P(\mathcal{X}, \mathbf{Z}, \Theta) = P(\mathcal{X} | \mathbf{Z}, \Theta) \cdot P(\mathbf{Z} | \Theta) \cdot P(\Theta), \tag{3.2}$$

where

$$P(\mathcal{X} | \mathbf{Z}, \Theta) = \prod_{u \in \mathcal{U}} \prod_{k=1}^K P(u | \Theta_k)^{z_{u,k}} \qquad P(\mathbf{Z} | \Theta) = \prod_{u \in V} \prod_{k=1}^K \pi_k^{z_{u,k}},$$

and $P(\Theta)$ represents the prior for the parameter set Θ. Figueredo and Jain [57] provide a general framework to rewrite the expectation-maximization algorithm in a way that automatically detects

---

[1]Improper priors are measures of uncertainty that do not necessarily represent probability densities [97]. For example, they can be used to express a preference for structural properties of the parameter under analysis, for which infinitely many values are equally likely.

an estimation of the most likely number of parameters. The key aspect is the modeling of the prior for the multinomial $\pi_1, \ldots \pi_K$: we can assume a large number K of components, and accept a k-th component only if it is supported by a sufficient number of observations to justify the "cost" of Θ_k. This can be encoded by penalizing the components on the basis of the complexity of the associated parameter set Θ_k:

$$P(\Theta) \propto \prod_{k=1}^{K} P(\Theta_k)\pi_k^{-\frac{N_k}{2}},$$

where $P(\Theta_k)$ is any specific prior on the parameter set of the k-th factor, the term $\prod_k \pi_k^{-\frac{N_k}{2}}$ represents an improper prior over parameters π_k and, finally, $N_k = |\Theta_k|$. In the standard EM framework (see Appendix A.1), the complete-data log likelihood can be expressed as

$$\mathcal{Q}(\Theta; \Theta') = \sum_{\mathbf{Z}} P(\mathbf{Z}|\mathcal{X}, \Theta') \log P(\mathcal{X}, \mathbf{Z}, \Theta) + \log P(\Theta)$$

$$\propto \sum_{u \in V} \sum_{k=1}^{K} \gamma_{u,k} \left\{ \log P(u|\Theta_k) + \log \pi_k \right\} - \sum_{k=1}^{K} \frac{N_k}{2} \log \pi_k + \sum_{k=1}^{K} \log P(\Theta_k).$$

Here, the term $\sum_{k=1}^{K} \log P(\Theta_k)$ acts as a regularization term for the model parameters Θ_k. Furthermore, the term $\sum_{k=1}^{K} N_k \log \pi_k$ enables a regularization on the multinomial parameters π_k: optimizing $\mathcal{Q}(\Theta; \Theta')$ with respect the latter yields the updated equation

$$\pi_k = \frac{\max\left\{0, \sum_u \gamma_{u,k} - N_k/2\right\}}{\sum_{k=1}^{K} \max\left\{0, \sum_u \gamma_{u,k} - N_k/2\right\}}. \tag{3.3}$$

The effect is that, within the EM algorithm, the estimation of the π_k parameters is adjusted and some components can be "annihilated": whenever a factor is not supported by a sufficient number of observations anymore, it is suppressed.

Since the optimal number of factors can be automatically detected, we can start with an arbitrary large initial value K, and then infer the final number K^* by looking at the components with non-zero priors. This is a general approach, which can be applied to any EM-based estimation procedure.

3.2   LATENT DIRICHLET ALLOCATION

We have seen that both ML and MAP provide estimations of the parameter set, and the difference between the two is given by the objective function: ML optimizes the likelihood $P(\mathcal{X}|\Theta)$, whereas MAP optimizes the posterior $P(\Theta|\mathcal{X})$. The strength of MAP over ML stems from the explicit account of the prior $P(\Theta)$. A full Bayesian approach focuses on a different perspective, by directly modeling the posterior distribution over the parameter set Θ:

$$P(\Theta|\mathcal{X}) = \frac{P(\mathcal{X}|\Theta)P(\Theta)}{P(\mathcal{X})}.$$

Table 3.1: Summary of notation

Model	SYMBOL	DESCRIPTION
	M	number of users
	N	number of items
	$\mathbf{R}$	$M \times N$ Rating Matrix
	n	number of observed preference in $\mathbf{R}$
	$\mathcal{X}$	Observations extracted from $\mathbf{R}$, i.e., $\mathcal{X} = \{x_1, \cdots, x_n\}$
	K	number of latent factors
	Θ	a generic parameter set to be estimated
	$x_m = \langle u, i, r \rangle$	the m-th observation in $\mathcal{X}$
LDA	θ	$\{\theta_u\}_{u \in \mathcal{U}}$
	θ_u	K-vector $\{\theta_{u,1}, \ldots \theta_{u,K}\}$: mixing proportion of factors for user u
	ϕ	$\{\phi_k\}_{k=1,\ldots,K}$
	ϕ_k	N-vector $\{\phi_{k,1}, \ldots, \phi_{k,N}\}$: mixing proportion of items for factor k
	z_m	latent factor associated with observation $m = \langle u, i \rangle$
	$\mathbf{Z}$	$M \times N$ matrix: user-factor assignments for each rating observation
	α	K- vector: Dirichlet priors over latent factors
	β	N-vector: Dirichlet priors over items
	$n_{u,i}^k$	number of preferences where user u and item i were associated with factor k (a dot generalizes over the corresponding dimension: for example $n_{u,\cdot}^k$ denotes the number of preferences where user u was associated with factor k)
	$\mathbf{n}_{u,i}^*$	$\{n_{u,i}^k\}_{k=1}^K$
	$\mathbf{n}_{u,*}^k$	$\{n_{u,i}^k\}_{i=1}^N$
URP, UCM	ϵ	$\{\epsilon_{k,i}\}_{k=1,\ldots,K, i \in \mathcal{I}}$
	$\epsilon_{k,i}$	V-vector $\{\epsilon_{k,i,1}, \ldots, \epsilon_{k,i,V}\}$: mixing proportion of ratings for factor k and item i
	γ	V- vector: Dirichlet priors over ratings
	$n_{u,i,r}^k$	number of observations $\langle u, i, r \rangle$ associated with factor k (a dot generalizes over the corresponding dimension)
Gaussian URP	$\mathbf{m}$	$M \times N$-matrix: $m_{u,i}$ represents the sum of the average ratings observed for user u and item i
	b	bias toward average ratings
	φ	$\{\varphi_{k,i}\}_{k \in \{1,\ldots,K\}, i \in \mathcal{I}}$
	$\varphi_{k,i}$	pair of Gaussian parameters $\{\mu_{k,i}, \sigma_{k,i}\}$
	η	variational Dirichlet parameters $\{\eta_u\}_{u \in \mathcal{U}}$ where $\eta_{,u} = \{\eta_{u,1}, \ldots, \eta_{u,k}\}$
	λ	variational multinomial parameters $\{\lambda_u,\}_{u \in \mathcal{U}}$, where $\lambda_u = \{\lambda_{u,1}, \ldots, \lambda_{u,k}\}$

We can see the difference with MAP here: estimating the posterior requires calculating the *evidence*

$$P(\mathcal{X}) = \int P(\mathcal{X}|\Theta) P(\Theta) \, d\Theta. \qquad (3.4)$$

Provided that the above quantities can be computed, it is possible to reformulate both the estimation and the prediction problems. In particular:

- Estimation can be accomplished by averaging over the whole parameter domain, $\hat{\Theta} = E[\Theta|\mathcal{X}] = \int \Theta P(\Theta|\mathcal{X}) \, d\Theta$, and

- Given $\tilde{x}$, we can compute its exact probability as

$$P(\tilde{x}|\mathcal{X}) = \int P(\tilde{x}|\Theta) P(\Theta|\mathcal{X}) \, d\Theta.$$

Let us consider again the likelihood of the pLSA model in Equation 2.9. The evidence can be expressed as:

$$P(\mathcal{X}) = \sum_{\mathbf{Z}} \int P(\mathcal{X}, \mathbf{Z}|\Theta) P(\Theta) \, d\Theta$$

$$= \int \left\{ \prod_{\langle u,i \rangle} \sum_{k} \theta_{u,k} \cdot \phi_{k,i} \right\} P(\Theta) \, d\Theta, \tag{3.5}$$

where $\mathbf{Z}$ is a $M \times N$ matrix that encodes the state of the latent variable for each observation $\langle u, i \rangle \in \mathcal{X}$, with $z_{u,i} \in \{1, \cdots, K\}$, and $P(\mathcal{X}, \mathbf{Z}|\Theta) = \prod_{\langle u,i \rangle} \theta_{u,z_{u,i}} \cdot \phi_{z_{u,i},i}$. Here, the latent parameters are averaged using a prior probability $P(\Theta)$. The tractability of the above integral (and, in general, of the integral Equation 3.4) is a central issue. Besides numerical methods, an alternative approach is to combine likelihood functions and prior distributions with convenient mathematical properties, including tractable analytic solutions to the integral. These families of prior distributions are known as *conjugate prior distributions*.

Recall that a conjugate prior $P(\Theta)$ of the likelihood $p(\mathcal{X}|\Theta)$ is a distribution that results in a posterior distribution, $P(\Theta|\mathcal{X})$ with the same functional form as the prior, but different parameters. In practice, the effect of likelihood on the posterior is only to "update" the prior parameters and not to change the prior's functional form.

The likelihood expressed by Equation 3.5, Θ consists of two parameter sets: $\{\theta_u\}_{u \in \mathcal{U}}$ and $\{\phi_k\}_{k=1,\ldots,K}$, where $\theta_u = \{\theta_{u,1}, \ldots \theta_{u,K}\}$ and $\phi_k = \{\phi_{k,1}, \ldots, \phi_{k,N}\}$ are multinomial distributions ranging over the latent factors and the items, respectively. A conjugate prior for a generic multinomial distribution $\mu = \{\mu_1, \ldots, \mu_n\}$ is the *Dirichlet Distribution* $Dir(\alpha)$ [31], parameterized by $\alpha = \{\alpha_1, \ldots, \alpha_n\}$ and defined as

$$Dir(\mu; \alpha) \triangleq P(\mu|\alpha) = \frac{1}{\Delta(\alpha)} \prod_{i=1}^{n} \mu_i^{\alpha_i - 1}; \qquad \Delta(\alpha) = \frac{\prod_{i=1}^{n} \Gamma(\alpha_i)}{\Gamma(\sum_{i=1}^{n} \alpha_i)}.$$

Exploiting the Dirichlet priors into Equation 3.5 yields

$$
\begin{aligned}
P(\mathcal{X}, \mathbf{Z}|\alpha, \beta) &= \int_{\Theta} \int_{\Phi} \prod_{\langle u,i \rangle} \theta_{u,k} \phi_{k,i} \, P(\theta_u|\alpha) P(\phi_k|\beta) \, \mathrm{d}\Theta \, \mathrm{d}\Phi \\
&= \int_{\Theta} \left(\prod_u \prod_k \theta_{u,k}^{n_{u,\cdot}^k} \right) \left(\prod_u \frac{1}{\Delta(\alpha)} \prod_k \theta_{u,k}^{\alpha_k - 1} \right) \mathrm{d}\Theta \\
&\quad \cdot \int_{\Phi} \left(\prod_k \prod_i \phi_{k,i}^{n_{\cdot,i}^k} \right) \left(\prod_k \frac{1}{\Delta(\beta)} \prod_i \phi_{k,i}^{\beta_i - 1} \mathrm{d} \right) \Phi \\
&= \prod_u \frac{1}{\Delta(\alpha)} \int_{\theta_u} \prod_k \theta_{u,k}^{n_{u,\cdot}^k + \alpha_k - 1} \, \mathrm{d}\theta_u \\
&\quad \cdot \prod_k \frac{1}{\Delta(\beta)} \int_{\phi_k} \prod_i \phi_{k,i}^{n_{\cdot,i}^k + \beta_i - 1} \, \mathrm{d}\phi_k \\
&= \prod_u \frac{\Delta(\alpha + \mathbf{n}_{u,\cdot}^*)}{\Delta(\alpha)} \cdot \prod_k \frac{\Delta(\beta + \mathbf{n}_{\cdot,*}^k)}{\Delta(\beta)},
\end{aligned}
\tag{3.6}
$$

where $\mathbf{n}_{u,\cdot}^*$ represents the vector $\{n_{u,\cdot}^k\}_{k=1,\dots,K}$ and, analogously, $\mathbf{n}_{\cdot,*}^k$ represents the vector $\{n_{\cdot,i}^k\}_{i \in \mathcal{I}}$. A summary of the notation adopted here is given in Table 3.1.

The above mathematical model has been proposed in [34] under the name *latent Dirichlet allocation* (LDA). Originally introduced for modeling text, it easily can be adapted to preference data. For example, in the case of implicit preferences, the generative process is characterized as follows:

1. For each latent factor $k = 1, \dots, K$ sample a multinomial distribution $\phi_k \sim Dir(\beta)$.

2. For each user $u \in \mathcal{U}$:

    (a) Sample the number n_u of item selections;

    (b) Choose $\theta_u \sim Dir(\alpha)$;

    (c) For each of the n_u items to be generated:

        i. Sample a topic $z \sim Disc(\theta_u)$;

        ii. Sample $i \sim Disc(\phi_z)$.

LDA is closely linked to pLSA, as it still captures the essence of explaining specific preference observations in terms of latent factors. However, the main differences lie in the role played by the prior distributions. The graphical representation in Figure 3.2 illustrates such differences. The adoption of the priors provides a better control of the parameters of the model, which are not fixed and are the result of a random process governed by α and β. As a result, inference and prediction can be approached by averaging over all possible distributions θ and ϕ. This ultimately

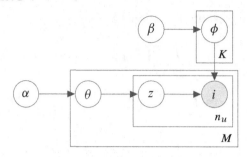

Figure 3.1: Graphical model for LDA.

results in better control over the sparsity of the data, and, hence, on overfitting issues, while still retaining the original expressive power of the pLSA modeling.

3.2.1   INFERENCE AND PARAMETER ESTIMATION

In a maximum likelihood perspective, the estimation phase consists of extracting the optimal α and β that maximize

$$P(\mathcal{X}|\alpha, \beta) = \sum_{\mathbf{Z}} \prod_u \frac{\Delta(\alpha + \mathbf{n}_{u,.}^*)}{\Delta(\alpha)} \cdot \prod_k \frac{\Delta(\beta + \mathbf{n}_{.,*}^k)}{\Delta(\beta)}.$$

Although latent Dirichlet allocation is still a relatively simple model, exact inference is generally intractable, due to the exponential number of possible factor assignments $\mathbf{Z}$. The solution to this is to use approximate inference algorithms, such as mean-field variational approximation [34], expectation propagation [134], and Gibbs sampling [75, 188]. We concentrate on Gibbs sampling and follow the presentation of [77]. The idea behind Gibbs sampling (a general introduction can be found in appendix A.3) is to create a random walk, or Markov process, that has the posterior $P(\mathbf{Z}|\mathcal{X})$ as its stationary distribution, and then to run the process long enough so that the resulting sample closely approximates a sample from the true $P(\mathbf{Z}|\mathcal{X})$ [62]. These samples can be used directly for parameter inference and prediction. In particular, they can be exploited in a stochastic variant of the EM algorithm where the E step consists in iteratively sampling $\mathbf{Z}$ using the Markov Chain, and the M step uses the sampled $\mathbf{Z}$ for estimating both the hyperparameters α and β (through ML), and the θ and ϕ (by averaging the corresponding posterior distributions).

Let us denote by m a pair $\langle u, i \rangle$ within $\mathcal{X}$. Then, $\mathcal{X}_{\neg m}$ (resp. $\mathbf{Z}_{\neg m}$) denotes the set $\mathcal{X}$ (resp. $\mathbf{Z}$) where the reference to $\langle u, i \rangle$ is ignored. Finally, let z_m denote the latent variable associated with the pair $\langle u, i \rangle$ in $\mathbf{Z}$. The intuition here is to define a Markov chain which approximates $P(\mathbf{Z}|\mathcal{X}, \alpha, \beta)$ by a repeated sampling over the components $P(z_m|\mathbf{Z}_{\neg m}, \mathcal{X}, \alpha, \beta)$, for each m.

We can observe the following.

$$
\begin{aligned}
P(z_m|\mathbf{Z}_{\neg m}, \mathcal{X}, \boldsymbol{\alpha}, \boldsymbol{\beta}) &= \frac{P(\mathbf{Z}, \mathcal{X}|\boldsymbol{\alpha}, \boldsymbol{\beta})}{P(\mathbf{Z}_{\neg m}, \mathcal{X}|\boldsymbol{\alpha}, \boldsymbol{\beta})} \\
&= \frac{P(\mathbf{Z}, \mathcal{X}|\boldsymbol{\alpha}, \boldsymbol{\beta})}{P(\mathbf{Z}_{\neg m}, \mathcal{X}_{\neg m}, x_m|\boldsymbol{\alpha}, \boldsymbol{\beta})} \\
&= \frac{P(\mathbf{Z}, \mathcal{X}|\boldsymbol{\alpha}, \boldsymbol{\beta})}{P(\mathbf{Z}_{\neg m}, \mathcal{X}_{\neg m}|\boldsymbol{\alpha}, \boldsymbol{\beta}) P(x_m|\boldsymbol{\alpha}, \boldsymbol{\beta})} \\
&\propto \frac{P(\mathbf{Z}, \mathcal{X}|\boldsymbol{\alpha}, \boldsymbol{\beta})}{P(\mathbf{Z}_{\neg m}, \mathcal{X}_{\neg m}|\boldsymbol{\alpha}, \boldsymbol{\beta})}.
\end{aligned}
\tag{3.7}
$$

Exploiting Equation 3.6, and by algebraic manipulations, we obtain the sampling equation

$$
P(z_m = k|\mathbf{Z}_{\neg m}, \mathcal{X}, \boldsymbol{\alpha}, \boldsymbol{\beta}) \propto \left(n_{u,\cdot}^k + \alpha_k - 1\right) \cdot \frac{n_{\cdot,i}^k + \beta_i - 1}{\sum_j \left(n_{\cdot,j}^k + \beta_j\right) - 1},
\tag{3.8}
$$

which defines the E step. As for the M step, we notice that the predictive distributions $\boldsymbol{\theta}$ and $\boldsymbol{\phi}$ can be characterized as

$$
\begin{aligned}
P(\boldsymbol{\theta}_u|\mathbf{z}_u, \boldsymbol{\alpha}) &\propto P(\mathbf{z}_u|\boldsymbol{\theta}_u) P(\boldsymbol{\theta}_u|\boldsymbol{\alpha}) = Dir(\boldsymbol{\alpha} + \mathbf{n}_{u,\cdot}^*) \\
P(\boldsymbol{\phi}_k|\mathbf{Z}, \mathcal{X}, \boldsymbol{\beta}) &\propto P(\mathcal{X}|\boldsymbol{\phi}_k, \mathbf{Z}) P(\boldsymbol{\phi}_k|\boldsymbol{\beta}) = Dir(\boldsymbol{\beta} + \mathbf{n}_{\cdot,*}^k),
\end{aligned}
$$

for which we can infer the mean values

$$
\theta_{u,k} = \frac{n_{u,\cdot}^k + \alpha_k}{\sum_{k'} \left(n_{u,\cdot}^{k'} + \alpha_{k'}\right)}
\tag{3.9}
$$

$$
\phi_{k,i} = \frac{n_{\cdot,i}^k + \beta_i}{\sum_j \left(n_{\cdot,j}^k + \beta_j\right)}.
\tag{3.10}
$$

Also since $\mathbf{Z}$ is known, we have

$$
\begin{aligned}
\log P(\mathcal{X}, \mathbf{Z}|\boldsymbol{\alpha}, \boldsymbol{\beta}) = &\sum_u \sum_{k=1}^K \left(\log \Gamma\left(\alpha_k + n_{u,\cdot}^k\right) - \log \Gamma\left(\alpha_k\right)\right) \\
&- \sum_u \left(\log \Gamma\left(n_u + \sum_{k=1}^K \alpha_k\right) - \log \Gamma\left(\sum_{k=1}^K \alpha_k\right)\right) \\
&+ \sum_k \sum_i \left(\log \Gamma\left(\beta_i + n_{\cdot,i}^k\right) - \log \Gamma\left(\beta_i\right)\right) \\
&- \sum_k \left(\log \Gamma\left(n^k + \sum_i \beta_i\right) - \log \Gamma\left(\sum_i \beta_i\right)\right).
\end{aligned}
\tag{3.11}
$$

Optimizing the above likelihood with respect to α, β requires numerical techniques. For example, the optimal value of α can be obtained through the fixed-point iteration [135]:

$$\alpha_k^{(t+1)} = \alpha_k^{(t)} \frac{\Psi\left(\alpha_k^{(t)} + n_{u,..}^k\right) - \psi\left(\alpha_k^{(t)}\right)}{\sum_u \left(\Psi\left(n_u + \sum_{k=1}^K \alpha_k^{(t)}\right) - \Psi\left(\sum_{k=1}^K \alpha_k^{(t)}\right)\right)}, \qquad (3.12)$$

where Ψ is the derivative logarithmic of the gamma function. A pseudo-code for the overall learning scheme is given in Algorithm 3.

Algorithm 3 Stochastic EM based on Gibbs sampling.

Require: The sets $\mathcal{U} = \{u_1, \ldots, u_M\}$ and $\mathcal{I} = \{i_1, \ldots, i_N\}$
        the rating observations $\mathcal{X}$, the number of latent topics K, initial hyperparameters α and β.
 1: *initializeTopicAssignments()* {Randomly assign topics}
 2: *iteration* $\leftarrow$ 0
 3: *converged* $\leftarrow$ *false*
 4: **while** *iteration* $<$ *nMaxIterations* **and** $\neg$*converged* **do**
 5:     **for all** $\langle u, i, r \rangle \in \mathbf{R}$ **do**
 6:         $z'_{u,i} \leftarrow$ *sampleTopic*(u, i, r) {According to Equation 3.7};
 7:         update counts using the new topic for the observation $\langle u, i, r \rangle$
 8:     **end for**
 9:     *updateHyperParams()* {By exploiting Equation 3.12}
10:     **if** (*iteration* $>$ *burn in*) **and** (*iteration%sample lag* $= 0$) **then**
11:         *sampleUserTopicsMixingProbabilities()* {According to Equation 3.9};
12:         *sampleItemSelectionProbabilities()* {According to Equation 3.10 };
13:         *converged* $\leftarrow$ *checkConvergence()*
14:     **end if**
15:     *iteration* $\leftarrow$ *iteration* $+ 1$
16: **end while**

3.2.2   BAYESIAN TOPIC MODELS FOR RECOMMENDATION

The basic LDA model discussed so far essentially provides support for implicit preferences. In particular, we can predict the preference for a pair $\langle u, i \rangle$ by relying on the hyperparameters,

$$P(i|u, \alpha, \beta) = \sum_k \int \phi_{k,i} \cdot P(\phi_k|\beta) \, d\phi_k \cdot \int \theta_{u,k} \cdot P(\theta_u|\alpha) \, d\theta_u$$

$$= \sum_k \frac{\Delta(\beta + \mathbf{e}^i)}{\Delta(\beta)} \frac{\Delta(\alpha + \mathbf{e}^k)}{\Delta(\alpha)},$$

where $\mathbf{e}^n$ is a vector where all $e_j^n = 1$ if $j = n$, and 0 otherwise. Alternatively, we can provide the estimation through the inferred predictive distributions θ and ϕ defined in Equations 3.9 and 3.10:

$$P(i|u, \theta, \phi) = \sum_k \theta_{u,k} \cdot \phi_{k,i}.$$

It is possible to reformulate the LDA for explicit preferences. The *User Rating Profile* (URP [123]) focuses on forced prediction and it is represented in Figure 3.2.

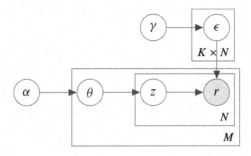

Figure 3.2: Graphical model for URP.

In this model, $\epsilon_{k,i}$ represents a multinomial distribution over the rating values in $\{1, \ldots, V\}$, corresponding to a latent factor k and an item i. The likelihood can be expressed as

$$P(\mathcal{X}|\boldsymbol{\alpha}, \boldsymbol{\gamma}) = \int \left\{ \prod_{\langle u,i,r \rangle} \sum_k \theta_{u,k} \epsilon_{k,i,r} \right\} P(\Theta|\boldsymbol{\alpha}, \boldsymbol{\gamma}) \, d\Theta,$$

for which a stochastic EM procedure similar to the one described by algorithm 3 can be devised (see [16] for details). Also, the probability of a rating is given by

$$P(r|u, i, \boldsymbol{\theta}, \boldsymbol{\epsilon}) = \sum_k \theta_{u,k} \cdot \epsilon_{k,i,r}, \tag{3.13}$$

or, alternatively, by solely relying on the $\boldsymbol{\alpha}$ and $\boldsymbol{\gamma}$ hyperparameters.

Barbieri et al. further develop this line of research [17] by proposing a Bayesian reformulation of the UCM model shown in Chapter 2, which explicitly models free prediction in a Bayesian setting. *Bayesian UCM* relies on a generative process, which takes into account both item selection and rating emission. Each user is modeled as a random mixture of topics, where the individual topic is then characterized both by a distribution modeling item-popularity within the considered user-community and by a distribution over preference values for those items. Thus, a user may be pushed to experience a certain item because she belongs to a community in which the category of that item occurs with a high probability, although this has no impact on the rating assigned to the aforesaid item category. The probability of observing an item is independent from the rating assigned, given the state of the latent variables.

The generative process behind the Bayesian UCM can be summarized as follows:

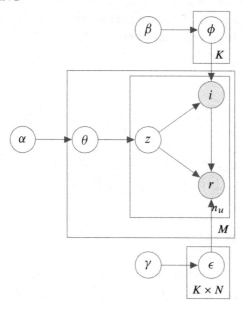

Figure 3.3: Graphical model for BUCM.

1. For each latent factor $z \in \{1, \cdots, K\}$:

    (a) Sample item selection components $\phi_z \sim Dir(\beta)$;

    (b) For each item $i \in \mathcal{I}$ sample rating probabilities $\varepsilon_{z,i} \sim Dir(\gamma)$.

2. For each user $u \in \mathcal{U}$:

    (a) Sample user community-mixture components $\theta_u \sim Dir(\alpha)$;

    (b) Sample the number of items n_u for the user u;

    (c) For each of the n_u items to select:

        i.   Sample a latent factor $z \sim Disc(\theta_u)$;

        ii. Choose an item $i \sim Disc(\phi_z)$;

        iii. Generate a rating value $r \sim Disc(\varepsilon_{z,i})$.

The corresponding joint likelihood can be expressed by the graphical model in Figure 3.3. Again, the adaptation of the stochastic EM framework for learning the parameters is relatively simple. The sampling equation is adapted to embody counters on both items and ratings,

$$p(z_m = k | \mathbf{Z}_{\neg m}, \mathcal{X}) \propto \left(n_{u,\cdot,\cdot}^k + \alpha_k - 1 \right) \cdot \frac{n_{\cdot,i,\cdot}^k + \beta_i - 1}{\sum_{i'=1}^{N} \left(n_{\cdot,i',\cdot}^k + \beta_{i'} \right) - 1} \cdot \frac{n_{\cdot,i,r}^k + \gamma_r - 1}{\sum_{r'=1}^{V} (n_{\cdot,i,r'}^k + \gamma_{r'}) - 1},$$

whereas the predictive distributions in the M step now also include the multinomial probability for a rating

$$\theta_{u,k} = \frac{n_{u,\cdot,\cdot}^k + \alpha_k}{n_u + \sum_{k=1}^{K} \alpha_k}; \qquad \phi_{i,k} = \frac{n_{\cdot,i,\cdot}^k + \beta_i}{\sum_{i=1}^{N} n_{\cdot,i,\cdot}^k + \beta_i}; \qquad \epsilon_{k,i,r} = \frac{n_{\cdot,ir}^k + \gamma_r}{\sum_{r'=1}^{V} n_{\cdot,ir'}^k + \gamma_{r'}}.$$

So far, all the proposed models represent extensions of the LDA basic model, and ratings were modeled as discrete variables associated with a multinomial distribution μ. Continuous ratings modeled through Gaussian distributions can be accommodated as well. Assume that, for a given latent factor k and an item i, we can devise the parameter set $\phi_{k,i} = \{\mu_{k,i}, \sigma_{k,i}\}$. Then, preference observations can be deemed as the result of the a process similar to the URP process described before, as shown in Figure 3.4. Within this graphical model, user and item biases (in particular, the baselines discussed in Section 1.4) can be made explicit to influence the rating, so that the term $m_{u,i}$ represents a specific bias for user u and item i, and the probability of a rating can be expressed as

$$P(r | u, i, \varphi_{k,i}, b) = \mathcal{N}\left(r; \mu_{k,i} + b \cdot m_{u,i}, \sigma_{k,i}\right).$$

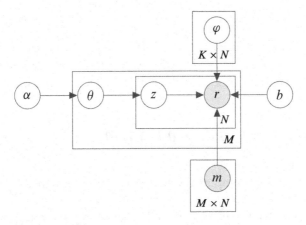

Figure 3.4: Graphical model for Gaussian URP.

Inference and parameter estimation of this model still can be approached through stochastic EM based on sampling. It is worth it to illustrate a different approach based on variational approximation. Let us consider the likelihood

$$P(\mathcal{X}|\alpha, b, \varphi) = \int \prod_u Dir(\theta_u|\alpha) \prod_{\langle u,i,r \rangle} \sum_k \theta_{u,k} P(r|u, i, \varphi_{k,i}, b) \, d\theta_u.$$

The idea behind variational inference is to provide an approximation $Q(\mathbf{Z}, \theta)$ to the posterior probability $P(\mathbf{Z}, \theta|\mathcal{X}, \varphi, b)$, which can be expressed in simpler mathematical forms. In our case, Q can be factorized into smaller components,

$$Q(\mathbf{Z}, \theta|\eta, \lambda) = \prod_u q(\theta_u|\eta_u) q(\mathbf{z}_u|\lambda_u),$$

where η and λ represent the set of variational parameters of Q, and, in particular, η represents the parameter of a Dirichlet distribution ranging over users, whereas λ represents a multinomial distribution.

Let us denote the parameters $\{\alpha, b, \varphi\}$ by $\mathcal{P}$, and the variational parameters $\{\eta, \lambda\}$ by $\mathcal{P}'$. It can be shown (see Appendix A.2) that Q enables the lower bound

$$\log P(\mathcal{X}|\mathcal{P}) \geq \mathcal{L}(\mathcal{P}, \mathcal{P}'),$$

where

$$\mathcal{L}(\mathcal{P}, \mathcal{P}') = \sum_{\mathbf{Z}} \int Q(\mathbf{Z}, \theta|\mathcal{P}') \log P(\mathcal{X}, \mathbf{Z}, \theta|\mathcal{P}) \, d\theta$$

$$+ \sum_{\mathbf{Z}} \int Q(\mathbf{Z}, \theta|\mathcal{P}') \log Q(\mathbf{Z}, \theta|\mathcal{P}') \, d\theta.$$

That is to say, $\mathcal{L}(\mathcal{P}, \mathcal{P}')$ can be used to approximate the log likelihood. For a fixed $\mathcal{P}$, the bound can be made tight by optimizing $\mathcal{L}(\mathcal{P}, \mathcal{P}')$ with respect to $\mathcal{P}'$, or, alternatively, with respect to $\mathcal{P}$. It can be shown (see Appendix A.2) that the optimal value for λ_u can be obtained by imposing the equality

$$\log q(\mathbf{z}_u|\lambda_u) = \sum_{\mathbf{Z}/\mathbf{z}_u} \int \log P(\mathcal{X}, \mathbf{Z}, \theta|\mathcal{P}) \prod_u q(\theta_u|\eta_u) \prod_{u' \neq u} q(\mathbf{z}_{u'}|\lambda_{u'}) \, d\theta + Const,$$

and solving for λ_u. Similarly, an optimal value for η_u can be obtained by imposing

$$\log q(\theta_u|\eta_u) = \sum_{\mathbf{Z}} \int \log P(\mathcal{X}, \mathbf{Z}|\mathcal{P}) \prod_{u' \neq u} q(\theta_{u'}|\eta_{u'}) \prod_u q(\mathbf{z}_u|\lambda_u) \, d\theta_{u' \neq u} + Const,$$

and solving for η_u. Notably, the integral can be simplified into a function of the parameters in $\mathcal{P}$. This creates circular dependencies between $\mathcal{P}$ and $\mathcal{P}'$, which naturally suggests an iterative algorithm much like EM, were the parameters are computed in turn by first estimating λ and η

as shown above, and then by exploiting the estimated value to optimize $\mathcal{L}(\mathcal{P}, \mathcal{P}')$ with respect to $\mathcal{P}$.

Notably, the variational hyperparameters λ can be interpreted as cluster membership and, combined with the estimated parameters ϕ and b, they can be exploited for prediction: given a pair u, i we can approximate the probability of rating r as

$$P(r|u, i, \lambda, \varphi, b) = \sum_{k=1}^{K} \lambda_{u,k} \cdot \mathcal{N}(r; \mu_{k,i} + b\, m_{u,i}, \sigma_{k,i}). \tag{3.14}$$

Variational inference and Gibbs sampling represent different approaches to solve the inference and estimation problem with full Bayesian models. Variational inference is a deterministic optimization method that is based on approximating the posterior distribution through a surrogate function, which is typically simpler. In practice, the time for each iteration of variational inference can be high, but, typically, the inference procedure converges fast. Also, the quality of the result strongly depends on the accuracy of the surrogate in approximating the true posterior. In fact, they can produce inaccurate results with overly simple approximations to the posterior.

By contrast, Gibbs sampling is based on statistical simulation. The Bayesian posterior distribution is computed here by sampling a reasonably large number of sampling points. Summing of the resulting posterior distribution values yields an approximation of the true data likelihood that asymptotically converge to the exact posterior. When the underlying distributions admit conjugate priors, the sampling step is relatively fast and simple to devise. However, for the approximation to be reliable, the number of required sampling iterations is typically high and the overall procedure is computationally demanding.

3.3 BAYESIAN CO-CLUSTERING

Collaborative filtering data exhibit global patterns (i.e., tendencies of some products to be "universally" appreciated) as well as significant local patterns (i.e., tendency of users belonging to a specific community to express similar preference indicators on the same items). Local preferences affect the performance of RS especially when the number of users and items grows, and their importance has been acknowledged by the current CF literature [161]. The interplay between local and global patterns is the reason why two users might agree perfectly on one topic while disagree completely on another.

We have seen in Section 2.2.2 that local patterns can be better detected by means of co-clustering approaches. Unlike traditional CF techniques, which try to discover similarities between users or items using clustering techniques or matrix decomposition methods, co-clustering approaches aim to partition data into homogenous blocks, enforcing a simultaneous clustering on both dimensions of the preference data.

It is natural to ask whether co-clustering relationships can be modeled in a Bayesian framework. And, in fact, several approaches that extend the basic LDA framework have been proposed

Table 3.2: Summary of notation

Model	SYMBOL	DESCRIPTION
Bi-LDA	K	number of latent factors involving users
	L	number of latent factors involving items
	ψ	$\{\psi_i\}_{i \in \mathcal{I}}$
	ψ_i	K-vector $\{\psi_{i,1}, \ldots \psi_{i,K}\}$: mixing proportion of factors for item i
	ϵ	$\{\epsilon_{k,l}\}_{k=1,\ldots,K, l=1,\ldots,L}$
	$\epsilon_{k,l}$	V-vector $\{\epsilon_{k,i,1}, \ldots, \epsilon_{k,i,V}\}$: mixing proportion of ratings for factor k and item i
	z_m	user latent factor associated with observation $m = \langle u, i \rangle$
	w_m	item latent factor associated with observation $m = \langle u, i \rangle$
	$\mathbf{Z}$	$M \times N$ matrix: user-factor assignments for each rating observation
	$\mathbf{W}$	$M \times N$ matrix: item-factor assignments for each rating observation
	α^1, α^2	K- vectors: Dirichlet priors over latent factors
	γ	V-vector: Dirichlet priors over ratings
RBC	$\mathbf{m}$	$M \times N$-matrix: $m_{u,i}$ represents the sum of the average ratings
	b	bias toward average ratings
	φ	$\{\varphi_{k,l}\}_{k=1,\ldots,K, l=1\ldots,L}$
	$\varphi_{k,l}$	pair of Gaussian parameters $\{\mu_{k,l}, \sigma_{k,l}\}$
	η_1, η_2	variational Dirichlet parameters $\{\eta_{1,u}\}_{u \in \mathcal{U}}$ and $\{\eta_{2,i}\}_{i \in \mathcal{I}}$, where $\eta_{1,u} = \{\eta_{1,u,1}, \ldots, \eta_{1,u,k}\}$ and $\eta_{2,i} = \{\eta_{2,i,1}, \ldots, \eta_{2,i,k}\}$
	λ_1, λ_2	variational multinomial parameters $\{\lambda_{1,u,}\}_{u \in \mathcal{U}}$ and $\{\lambda_{2,i,}\}_{i \in \mathcal{I}}$, where $\lambda_{1,u} = \{\lambda_{1,u,1}, \ldots, \lambda_{1,u,k}\}$ and $\lambda_{2,i} = \{\lambda_{2,i,1}, \ldots, \eta_{2,i,k}\}$

[155, 178, 179, 196]. The underlying idea is similar to the one discussed in Section 2.2.2. That is, we assume two independent latent factors, namely z and w, where z models the intuition that users can be grouped into communities, whereas w represents items groupings into categories. Therefore, each preference observation is associated with a pair $\langle z, w \rangle$.

As an example, the *Bi-LDA* model, proposed in [155], extends the URP model by employing two interacting LDA models, thus enforcing a simultaneous clustering of users and items in homogeneous groups.

These relationships can be observed in the graphical models in Figure 3.5, which encodes the following generative process:

1. For each user $u \in \mathcal{U}$ sample $\boldsymbol{\theta}_u \sim Dir(\boldsymbol{\alpha}^1)$.

2. For each item $i \in \mathcal{I}$ sample $\boldsymbol{\psi}_i \sim Dir(\boldsymbol{\alpha}^2)$.

3. For each pair $(k, l) \in \{1, \ldots, K\} \times \{1, \ldots, L\}$ sample $\epsilon_{k,l} \sim Dir(\boldsymbol{\gamma})$.

4. For each pair $m = \langle u, i \rangle$:

   (a) sample $z_m \sim Disc(\boldsymbol{\theta}_u)$;

   (b) sample $w_m \sim Disc(\boldsymbol{\psi}_i)$;

   (c) sample $r \sim Disc(\epsilon_{z_m, w_m})$.

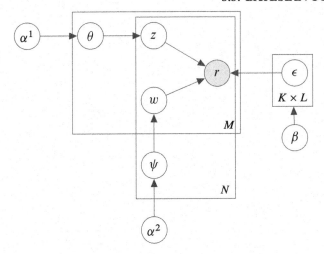

Figure 3.5: Graphical model for Bi-LDA.

Prediction can be devised by marginalizing on both user and item factors, by means of the estimated predictive distributions:

$$P(r|u,i,\epsilon,\boldsymbol{\theta},\boldsymbol{\psi}) = \sum_k \sum_k \epsilon_{k,l,r} \cdot \theta_{u,k} \cdot \psi_{i,l}. \tag{3.15}$$

Continuous modeling of explicit preferences can be accomplished in a similar way. The *residual Bayesian co-clustering (RBC)* proposed in [181] extends the Gaussian URP model, discussed in the previous section, to include latent factors on both the user and the item dimensions. The model is represented graphically in Figure 3.6. Within the model, the Gaussian parameters can be specified on the user and item latent factors, i.e., $\varphi_{k,l} \triangleq \{\mu_{k,l}, \sigma_{k,l}\}$, and the probability of observing a rating r for a pair $\langle u,i \rangle$ is given by

$$P(r|u,i,\varphi_{k,l},b) = \mathcal{N}(r; \mu_{k,l} + b \cdot m_{u,i}, \sigma_{k,l}).$$

The main difference with the model shown in Figure 3.4 is the introduction of a latent variable on items. As a consequence, inference and parameter estimation can be easily adapted: in [181], a variational EM algorithm is proposed by introducing the variational distribution

$$Q(\mathbf{Z}, \mathbf{W}, \boldsymbol{\theta}, \boldsymbol{\psi}|\mathcal{P}') = \left(\prod_u q(\boldsymbol{\theta}_u|\boldsymbol{\eta}_{1,u})\right)\left(\prod_i q(\boldsymbol{\psi}_i|\boldsymbol{\eta}_{2,i})\right) \cdot \left(\prod_{u,i} q(z_{u,i}|\boldsymbol{\lambda}_{1,u})q(w_{u,i}|\boldsymbol{\lambda}_{2,i})\right),$$

where $\mathcal{P}' = \{\boldsymbol{\eta}_1, \boldsymbol{\eta}_2, \boldsymbol{\lambda}_1, \boldsymbol{\lambda}_2\}$ represents the set of variational parameters of Q, $\boldsymbol{\eta}_1, \boldsymbol{\eta}_2$ represents the parameters of a Dirichlet distribution, and $\boldsymbol{\lambda}_1, \boldsymbol{\lambda}_2$ represents parameters of a multinomial

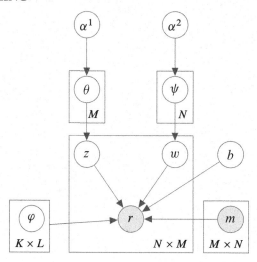

Figure 3.6: Graphical model for RBC.

distribution, both ranging over users and items respectively. These variational parameters, together with the model parameters ϕ and b, are learned by the EM procedure and can be exploited in the prediction equation directly adapted from Equation 3.14:

$$P(r|u, i, \lambda_1, \lambda_2, \varphi, b) = \sum_{k=1}^{K} \sum_{l=1}^{L} \lambda_{1,u,k} \lambda_{2,i,l} \mathcal{N}(r; \mu_{k,l} + b \cdot m_{u,i}).$$

3.3.1 HIERARCHICAL MODELS

A major weakness of the current approaches to co-clustering is the static structure enforced by fixed row/column blocks where both users and items have to fit. For example, the movies "Titanic" and "Avatar" are typically associated with different categories: the former is about romance, whereas the latter can be considered an action, sci-fi movie. Assuming a global and unique partition on the item-set, we can expect to see the movies in different partitions. However, that structure would fail to recognize a group of users who are fans of James Cameron, the director for both the movies. Analogously, any method associating the two movies with the same partition would fail in identifying the difference in genre.

The issue in the previous example is that different user groups can infer different interpretations of item categories. A more flexible structure, where item categories are conditioned by user communities, would better model such situations, e.g., by allowing "Titanic" and "Avatar" to be observed in the same item category within the "Cameron fans" group, and in different categories outside. Notice that traditional clustering approaches are not affected by this problem, as they only concentrate on local patterns in one dimension of the rating matrix. The drawback, however,

Table 3.3: Summary of notation

Model	SYMBOL	DESCRIPTION
BH-Forced	ψ	parameter set $\{\psi_{i,k}\}_{k=1,\dots,K, i \in \mathcal{I}}$
	$\psi_{i,k}$	L-vector: mixing proportion for the item category l and the user-topic k
	ϵ	parameter set $\epsilon_{k,l}$
	$\epsilon_{k,l}$	V-vector: distribution over rating values for the co-cluster k, l
	α^1, α^2	K- vector: Dirichlet priors on user communities and item categories
	γ	V-vector: Dirichlet priors on rating values
BH-Free	ψ	parameter set $\{\psi_k\}_{k=1,\dots,K}$
	ψ_k	L-vector: mixing proportion for the user-topic k
	ϕ	parameter set $\phi_{k,l}$
	$\phi_{k,l}$	N-vector: mixing proportion for each item i in the co-cluster k, l
	β	N-vector: Dirichlet priors on items
	$n_{u,i,r}^{k,l}$	number of times user u and item i are associated with rating r corresponding to latent factors k, l (a dot generalizes over the corresponding dimension)
	$\mathbf{n}_{u,\cdot,\cdot}^{*,\cdot}$	$\{n_{u,\cdot,\cdot}^{k,\cdot}\}_{k=1,\dots,K}$
	$\mathbf{n}_{\cdot,\cdot,\cdot}^{k,*}$	$\{n_{\cdot,\cdot,\cdot}^{k,l}\}_{l=1,\dots,L}$
	$\mathbf{n}_{\cdot,\cdot,*}^{k,l}$	$\{n_{\cdot,\cdot,r}^{k,l}\}_{r=1,\dots,V}$
	$\mathbf{n}_{\cdot,*,\cdot}^{k,l}$	$\{n_{\cdot,i,\cdot}^{k,l}\}_{i \in \mathcal{I}}$

is that they ignore structural information in the other dimension, which by the converse can be exploited both for more accurate prediction and user profiling.

Specific groups of users tend to be co-related according to different subsets of features. However, though semantically-related, two users with (possibly several) differences in their preferences would hardly be recognized as actually similar by any global model imposing a fixed structure for item categories. Individual users can be intended as a mixture of latent concepts, each of which being a suitable collection of characterizing features. Accordingly, two users are considered as actually similar if both represent at least a same concept. Viewed in this perspective, the identification of *local patterns*, i.e., of proper combinations of users and items, would lead to the discovery of natural clusters in the data, without incurring the aforesaid difficulties. Consider the toy example in Figure 3.7, where homogenous blocks exhibiting similar rating patterns are highlighted. There are seven users clustered in two main communities. Community 1 is characterized by three main topics (with groups $d_{11} = \{i_1, i_2, i_3\}$, $d_{12} = \{i_4, i_5, i_6, i_7\}$ and $d_{13} = \{i_8, i_9, i_{10}\}$), whereas community 2 includes four main topics (with groups $d_{21} = \{i_1, i_4, i_5\}$, $d_{22} = \{i_2, i_3, i_7\}$, $d_{23} = \{i_6, i_{10}\}$ and $d_{24} = \{i_8, i_9\}$). The difference with respect to traditional co-clustering techniques is that different communities group the same items differently. This introduces a topic hierarchy that, in principle, increases the semantic power of the overall model.

An extension of the Bayesian co-clustering framework to incorporate such ideas has been made in [20]. The key idea is that there exists a set of user communities, each one describing different tastes of users and their corresponding rating patterns. Each user community is then

		i_1	i_2	i_3	i_4	i_5	i_6	i_7	i_8	i_9	i_{10}
Community 1	u_1	1		1	5		4	5		2	2
	u_2	1	1			5	4	4	5	2	2
	u_3	1	1	1	4	5			5	2	
	u_4		1	1		5	4	5	4		2
Community 2	u_5	5		4	5	5	1	4	3		1
	u_6		4	4	5	5	1	4	3	3	1
	u_7	5	4		5		1	4	3	3	

$d_1 = \{i_1, i_2, i_3\}$
$d_2 = \{i_4, i_5, i_6, i_7\}$
$d_3 = \{i_9, i_{10}\}$

$d_1 = \{i_1, i_4, i_5\}$
$d_2 = \{i_2, i_3, i_7\}$
$d_3 = \{i_6, i_{10}\}$
$d_4 = \{i_8, i_9\}$

Figure 3.7: Example of local pattern in CF data.

modeled as a random mixture over latent topics, which can be interpreted as item-categories. Given a user u, we can foresee her preferences on a set of items $\mathcal{I}(u)$ by choosing an appropriate user community z and then choosing an item category w for each item in the list. The choice of the item category w actually depends on the selected user community z. Finally, the preference value is generated by considering the preference of users belonging to the group z on items of the category w.

A first coarse-grained generative process can be devised as an adaptation of the Bi-LDA model, and it is graphically depicted in Figure 3.8:

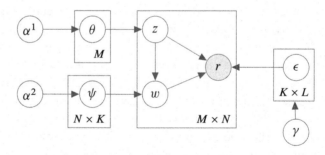

Figure 3.8: *BH-forced* generative model.

1. For each user $u \in \mathcal{U}$ sample $\boldsymbol{\theta}_u \sim Dir(\alpha^1)$.

2. For each item $i \in \mathcal{I}$ and user community $z \in \{1, \ldots, K\}$ sample the mixture components $\psi_{z,i} \sim Dir(\alpha^2)$.

3. For each topic $w \in \{1, \ldots, L\}$ and user community $z = \{1, \cdots, K\}$, sample rating probabilities $\epsilon_{z,w} \sim Dir(\gamma)$.

4. For each active pair $m = \langle u, i \rangle$:

   (a) Choose a user factor $z_m \sim Disc(\theta_u)$;

   (b) Choose a topic $w_m \sim Disc(\psi_{z_m,i})$;

   (c) Generate a rating value $r \sim Disc(\epsilon_{z_m,w_m})$.

We name this model *Bayesian hierarchical co-clustering focused on forced prediction (BH-Forced)*. Figure 3.9 provides a possible hierarchical model for the data in Figure 3.7, where we assume that $K = 2$ and $L = 4$. Then, (a) represents assignments z_m corresponding to each observed pair $m \equiv \langle u, i \rangle$, whereas (b) represents assignments w_m. Furthermore, (c) represents the probability θ, and (d) and (e) represent ψ for the two communities associated with users. By applying the generative process described above, the interested reader easily can verify that each observed rating can be explained as the result of two latent factors in hierarchical relationships. For example, let us consider the observation $\langle u_5, i_5 \rangle$. According to the devised generative process, we first pick user community 2 for u_5, exploiting table (c). Next, we assign item category 1 to item i_5, by drawing from the available categories according to the probability in table (e). Finally, given the co-cluster $\langle 2, 1 \rangle$, we observe rating 5 by picking randomly according to the related rating distribution in table (f).

The BH-forced approach focuses on forced-prediction, as it models the prediction of preference values for each observed user-item pair, and it does not explicitly take into account item selection. It is also possible to accommodate the model to provide explicit support to the item selection component. Figure 3.10 illustrates the *Bayesian Hierarchical focused on Free Prediction (BH-Free)*. Each user is modeled as a random mixture of topics, where the individual topic is then characterized both by a distribution modeling item-popularity within the considered user-community and by a distribution over preference values for those items. The distribution of items given the topic variable w depends on the choice of the user community: this enforces an explicit modeling of item popularity, both within a category and within a community, and hence provides a high degree of flexibility. The rating prediction components maintain almost the same structure as in the *BH-forced* model, and hence, even the accuracy is almost the same.

The role of item selection within the new *BH-free* model can be noticed in the underlying generative process:

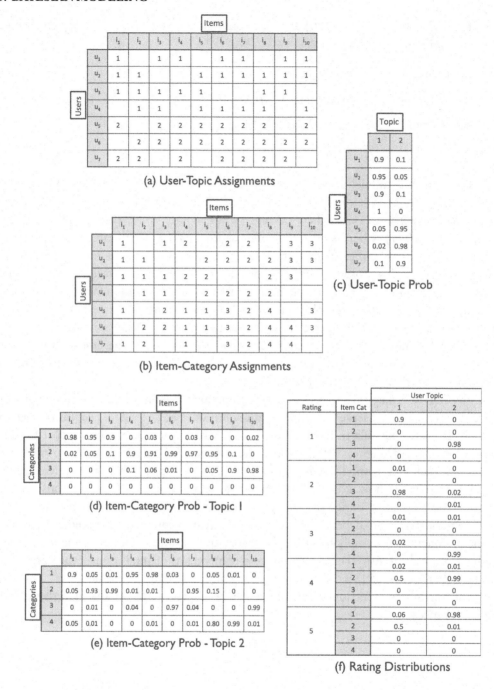

Figure 3.9: Probabilistic modeling of local patterns.

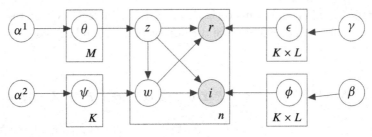

Figure 3.10: Generative model for BH-free.

1. For each user $u \in \mathcal{U}$ sample $\theta_u \sim Dir(\alpha^1)$.

2. For each user community $z \in \{1, \ldots, K\}$ sample the mixture components $\psi_z \sim Dir(\alpha^2)$.

3. For each topic $w \in \{1, \ldots, L\}$ and user community $z = \{1, \cdots, K\}$;

   (a) Sample item selection components $\phi_{z.w} \sim Dir(\beta)$;

   (b) Sample rating probabilities $\epsilon_{z,w} \sim Dir(\gamma)$.

4. For each $u \in \mathcal{U}$:

   (a) Sample the number of items n_u;

   (b) For $m \in \{1, \ldots, n_u\}$:

        i. Choose a user attitude $z_m \sim Disc(\theta_u)$;

        ii. Choose a topic $w_m \sim Disc(\psi_{z_m})$;

        iii. Choose an item $i \sim Disc(\phi_{z_m,w_m})$;

        iv. Generate a rating value $r \sim Disc(\epsilon_{z_m,w_m})$.

BH-free is aimed at inferring the tendency of a user to experience some items over others, independent of her/his rating values. The model assumes that this tendency is influenced by implicit and hidden factors that characterize each user community. This can be considered as an extension of the UCM model discussed in Sections 2.2.1 and 3.2.2, for two main reasons: first, it adds a hierarchical co-clustering structure, thus complying to the original idea of modeling local patterns; second, it accommodates a Bayesian modeling, which allows better control of data sparseness.

The inference process is similar for both BH-forced and BH-free. Concerning the latter model, there is a small overhead due to the explicit modeling of item selection. Hence, in the following, we only sketch the derivation of the sampling equations for this model. The equations for the forced version can be derived accordingly.

The notation used in our discussion is summarized in Table 3.3. Given the hyperparameters $\alpha^1, \alpha^1, \beta$, and γ, the joint distribution of the observations $\mathcal{X}$, and the assignments $\mathbf{Z}, \mathbf{W}$ can be factored as:

$$P(\mathcal{X}, \mathbf{Z}, \mathbf{W} | \alpha^1, \alpha^2, \beta, \gamma) = \int P(\mathbf{Z} | \theta) P(\theta | \alpha^1) \, d\theta \int P(\mathbf{W} | \mathbf{Z}, \psi) P(\psi | \alpha^2) \, d\psi$$
$$\cdot \int \int P(\mathcal{X} | \mathbf{Z}, \mathbf{W}, \phi, \epsilon) P(\phi | \beta) P(\epsilon | \gamma) \, d\phi \, d\theta.$$

By rearranging the components and grouping the conjugate distributions, the complete data likelihood can be expressed as:

$$P(\mathcal{X}, \mathbf{Z}, \mathbf{W} | \alpha^1, \alpha^2, \beta, \gamma) = \prod_{u=1}^{M} \frac{\Delta(n_{u,\cdot,\cdot}^{*,\cdot} + \alpha)}{\Delta(\alpha^1)} \cdot \prod_{k=1}^{K} \frac{\Delta(n_{\cdot,\cdot,\cdot}^{k,*} + \alpha^2)}{\Delta(\alpha^2)}$$
$$\cdot \prod_{k=1}^{K} \prod_{l=1}^{L} \frac{\Delta(n_{\cdot,\cdot,*}^{k,l} + \gamma)}{\Delta(\gamma)} \cdot \prod_{k=1}^{K} \prod_{l=1}^{L} \frac{\Delta(n_{\cdot,*,\cdot}^{k,l} + \beta)}{\Delta(\beta)}.$$

The latter is the starting point for the inference of all the topics underlying the generative process, as the posterior on $\mathbf{Z}, \mathbf{W}$ can be written as:

$$P(\mathbf{Z}, \mathbf{W} | \mathcal{X}, \alpha^1, \alpha^2, \gamma, \beta) = \frac{P(\mathbf{Z}, \mathbf{W}, \mathcal{X} | \alpha^1, \alpha^2, \gamma, \beta)}{P(\mathcal{X} | \alpha^1, \alpha^2, \gamma, \beta)}.$$

As already mentioned, this formula is intractable because of its denominator. We can resort again to Gibbs sampling by defining a Markov chain, in which, at each step, inference can be accomplished by exploiting the full conditional $P(z_m = k, w_m = l | \mathbf{Z}_{\neg m}, \mathbf{W}_{\neg m}, \mathcal{X}, \alpha^1, \alpha^2, \gamma, \beta)$. As usual, z_m (resp. w_m) is the assignment for the cell m of the matrix $\mathbf{Z}$ (resp. $\mathbf{W}$), and $\mathbf{Z}_{\neg m}$ ($\mathbf{W}_{\neg m}$) denotes the remaining latent factor assignments. The chain is hence defined by iterating over the available states m. The Gibbs sampling algorithm estimates the probability of assigning the pair k, l to the m-th observation $\langle u, i, r \rangle$, given the assignment corresponding to all other rating observations:

$$P(z_m = k, w_m = l | \mathbf{Z}_{\neg m}, \mathbf{W}_{\neg m}, \mathcal{X}, \alpha^1, \alpha^2, \gamma, \beta) \propto$$
$$\left(n_{u,\cdot,\cdot}^{k,\cdot} + \alpha_k^1 - 1 \right) \cdot \left(n_{\cdot,\cdot,\cdot}^{k,l} + \alpha_l^2 - 1 \right)$$
$$\cdot \frac{n_{\cdot,\cdot,r}^{k,l} + \gamma_r - 1}{\sum_{r'=1}^{V} (n_{\cdot,\cdot,r+}^{k,l} + \gamma_r) - 1} \cdot \frac{n_{\cdot,i,\cdot}^{k,l} + \beta_i - 1}{\sum_{i'=1}^{N} (n_{\cdot,i',\cdot}^{k,l} + \beta_i) - 1}. \tag{3.16}$$

The above equation specifies the E step in the Stochastic EM procedure. The M step provides an estimate of the model parameters:

$$\theta_{u,k} = \frac{n_{u,\cdot,\cdot}^{k,\cdot} + \alpha_k^1}{n_{u,\cdot,\cdot}^{\cdot,\cdot} + \sum_{k=1}^{K} \alpha_k^1}; \qquad \psi_{k,l} = \frac{n_{\cdot,\cdot,\cdot}^{k,l} + \alpha_l^2}{n_{\cdot,\cdot,\cdot}^{k,\cdot} + \sum_{l=1}^{L} \alpha_l^2};$$

$$\epsilon_{k,l,r} = \frac{n_{\cdot,\cdot,r}^{k,l} + \gamma_r}{n_{\cdot,\cdot,\cdot}^{k,l} + \sum_{r'=1}^{V} \gamma_{r'}}; \qquad \phi_{k,l,i} = \frac{n_{\cdot,i,\cdot}^{k,l} + \beta_i}{n_{\cdot,\cdot,\cdot}^{k,l} + \sum_{i'=1}^{N} \beta_{i'}},$$

and the hyperparameter can be estimated by means of numerical methods.

Finally, given the pair $\langle u, i \rangle$, we compute the probability of observing the rating value r in a free prediction context:

$$p(r, i | u, \boldsymbol{\theta}, \boldsymbol{\psi}, \boldsymbol{\phi}, \boldsymbol{\epsilon}) = \sum_{k=1}^{K} \sum_{l=1}^{L} \theta_{u,k} \cdot \psi_{k,l} \cdot \phi_{k,l,i} \cdot \epsilon_{k,l,r}. \qquad (3.17)$$

Notice the explicit reference, in Equation 3.17, to the $\phi_{k,l,i}$ component that models the probability of i being selected within co-cluster k, l. Such a component biases the ranking toward relevant items, thus providing the required adjustment that makes the model suitable for both prediction and recommendation accuracy. Compared to Equation 3.17, the prediction equation for BH-forced does not include such a component and relies on a different specification of the ψ distribution:

$$p(r | u, i, \boldsymbol{\theta}, \boldsymbol{\psi}, \boldsymbol{\epsilon}) = \sum_{k=1}^{K} \sum_{l=1}^{L} \theta_{u,k} \cdot \psi_{k,l,i} \cdot \epsilon_{k,l,r}. \qquad (3.18)$$

3.4   BAYESIAN MATRIX FACTORIZATION

We can finally review the PMF model discussed in Section 2.3.2 under a Bayesian perspective. We already have seen in Section 3.1 that the parameters $\mathbf{P}$ and $\mathbf{Q}$ are regularized using Gaussian priors governed by the variance hyperparameters. Thus, the optimal values for the parameters are obtained by MAP estimation.

We can take a step further and study more general Bayesian formulations for PMF. In Section 3.1, we initially assumed a constant variance for the Gaussian prior: this implies that the latent features are independent. We also discussed a more general formulation of the priors, through multivariate Gaussian governed by the $\mu_U, \mu_I, \Sigma_U, \Sigma_I$ hyperparameters. These general hyperparameters enable an adaptive regularization, as they can better model specific cases (e.g., users with less/more preferences) and dependencies among the latent factors. The general model is graphically represented in Figure 3.11.

Since $\mathbf{P}$ and $\mathbf{Q}$ depend on the hyperparameters $\mathcal{P} = \{\mu_U, \mu_I, \Sigma_U, \Sigma_I\}$, we can reformulate the inference and parameter estimation in a more general way, by marginalizing over all

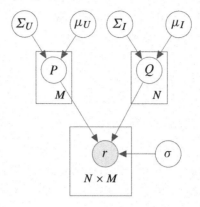

Figure 3.11: A generalized PMF model.

possible values of such parameters, given the hyperparameters:

$$P(\mathcal{X}|\mathcal{P},\sigma) = \int \int \prod_u P(\mathbf{P}_u|\boldsymbol{\mu}_U,\boldsymbol{\Sigma}_U) \prod_i P(\mathbf{Q}_i|\boldsymbol{\mu}_I,\boldsymbol{\Sigma}_I) \prod_{\langle u,i,r \rangle} P(r|\mathbf{P}_u^T\mathbf{Q}_i,\sigma)\, d\mathbf{P}\, d\mathbf{Q}. \quad (3.19)$$

Here, we change the focus with respect to the original PMF, by focusing on the optimal hyperparameter set $\mathcal{P}$, for which $\mathbf{P}$ and $\mathbf{Q}$ can be averaged in a Bayesian setting. This approach represents a generalization of the PMF, named *parametric PMF (PPMF)* and has been investigated extensively [91, 115, 180].

Inference and parameter estimation can be accomplished by means of a variational EM procedure, as follows. We can introduce a wider set of parameters $\mathcal{P}' = \{\boldsymbol{\lambda}_{1,u}, \boldsymbol{\nu}_{1,u}, \boldsymbol{\lambda}_{2,i}, \boldsymbol{\nu}_{2,i}\}_{u\in\mathcal{U},i\in\mathcal{I}}$, such that the posterior $P(\mathbf{P},\mathbf{Q}|\mathcal{P},\mathcal{X})$ can be approximated by the factorized variational distribution $Q(\mathbf{P},\mathbf{Q}|\mathcal{P}')$ defined as

$$Q(\mathbf{P},\mathbf{Q}|\mathcal{P}') = \prod_u q(\mathbf{P}_u|\boldsymbol{\lambda}_{1,u},\mathrm{diag}(\boldsymbol{\nu}_{1,u})) \prod_i q(\mathbf{Q}_i|\boldsymbol{\lambda}_{2,i},\mathrm{diag}(\boldsymbol{\nu}_{2,i})).$$

Here, the factor q represent a K-dimensional Gaussian distribution. Based on Q, we can express the lower bound

$$\mathcal{L}(\mathcal{P},\mathcal{P}') = \int Q(\mathbf{P},\mathbf{Q}|\mathcal{P}') \log P(\mathcal{X},\mathbf{P},\mathbf{Q}|\mathcal{P})\, d\mathbf{P}\, d\mathbf{Q}$$
$$- \int Q(\mathbf{P},\mathbf{Q}|\mathcal{P}') \log Q(\mathbf{P},\mathbf{Q}|\mathcal{P}')\, d\mathbf{P}\, d\mathbf{Q},$$

for the data likelihood, which enables the standard variational EM procedure where we alternatively optimize $\mathcal{L}(\mathcal{P},\mathcal{P}')$ with respect to $\mathcal{P}$ and $\mathcal{P}'$ (mathematical details can be devised in Appendix A.2). Also, since Q is factorized, closed forms for the optimal values for $\mathcal{P}'$ in the E step can be obtained by solving the equations

$$\log q(\mathbf{P}_u|\boldsymbol{\lambda}_{1,u}, \mathrm{diag}(\boldsymbol{\nu}_{1,u})) = \int Q(\mathbf{P}, \mathbf{Q}|\mathcal{P}') \log P(\mathcal{X}, \mathbf{P}, \mathbf{Q}|\mathcal{P}) \, d\mathbf{P}_{\neg \mathbf{u}} \, d\mathbf{Q} + \mathit{Const}$$

$$\log q(\mathbf{Q}_i|\boldsymbol{\lambda}_{2,i}, \mathrm{diag}(\boldsymbol{\nu}_{2,i})) = \int Q(\mathbf{P}, \mathbf{Q}|\mathcal{P}') \log P(\mathcal{X}, \mathbf{P}, \mathbf{Q}|\mathcal{P}) \, d\mathbf{P} \, d\mathbf{Q}_{\neg \mathbf{i}} + \mathit{Const},$$

for the given variables.

The variational parameters $\mathcal{P}'$ can also be employed for solving the prediction problem. In particular, since $Q(\mathbf{P}_u, \mathbf{Q}_i|\mathcal{P}')$ approximates the true posterior $P(\mathbf{P}_u, \mathbf{Q}_i|\mathcal{X}, \mathcal{P})$, we can derive

$$\begin{aligned} \{\hat{\mathbf{P}}_u, \hat{\mathbf{Q}}_i\} &= \underset{\mathbf{P}_u, \mathbf{Q}_i}{\mathrm{argmax}} \, P(\mathbf{P}_u, \mathbf{Q}_i|\mathcal{X}, \mathcal{P}) \\ &\approx \underset{\mathbf{P}_u, \mathbf{Q}_i}{\mathrm{argmax}} \, Q(\mathbf{P}_u, \mathbf{Q}_i|\mathcal{P}') \\ &= \{\boldsymbol{\lambda}_{1,u}, \boldsymbol{\lambda}_{2,i}\}, \end{aligned}$$

and hence

$$P(r|u, i, \mathcal{X}, \mathcal{P}) \approx P(r|u, i, \boldsymbol{\lambda}_{1,u}, \boldsymbol{\lambda}_{2,i}, \sigma) = \mathcal{N}(r; \boldsymbol{\lambda}_{1,u}^T \boldsymbol{\lambda}_{2,i}, \sigma).$$

An alternative, full Bayesian treatment includes a further model averaging on the $\mathcal{P}$ parameters [168]. We can introduce the hyperparameters $\Theta_0 \triangleq \{\boldsymbol{\mu}_0, \beta_0, \mathbf{W}_0, \eta_0\}$, which can be exploited in a Gaussian-Wishart distribution. The latter are known to be conjugate to the Gaussian multivariate distribution [31, Section 2.3.6], and is expressed as:

$$P(\boldsymbol{\mu}, \boldsymbol{\Sigma}|\Theta_0) = \mathcal{N}(\boldsymbol{\mu}|\boldsymbol{\mu}_0, \beta_0 \boldsymbol{\Sigma}) \mathcal{W}(\boldsymbol{\Sigma}|\mathbf{W}_0, \eta_0),$$

where $\mathcal{W}(\boldsymbol{\Sigma}|\mathbf{W}_0, \eta_0)$ represents the Wishart distribution with ν_0 degrees of freedom and scale $\mathbf{W}_0$. Then, by adopting the notation $\Theta_U \triangleq \{\boldsymbol{\mu}_U, \boldsymbol{\mu}_U\}$, $\Theta_I \triangleq \{\boldsymbol{\mu}_V, \boldsymbol{\mu}_V\}$, we can express the rating probability as

$$P(r|u, i, \mathcal{X}, \Theta_0) = \int P(r|\mathbf{P}_u, \mathbf{Q}_i, \sigma) \cdot P(\mathbf{P}_u, \mathbf{Q}_i|\mathcal{X}, \Theta_U, \Theta_I) \qquad (3.20)$$
$$\cdot P(\Theta_U, \Theta_I|\Theta_0) \, d\mathbf{P} \, d\mathbf{Q} \, d\Theta_U \, d\Theta_I.$$

Usually, Θ_0 is specified by means of constant values. For example, [168] suggests $\boldsymbol{\mu}_0 = 0$, $\nu_0 = K$ and $\mathbf{W}_0$ as the identity matrix. The dependencies among the variables of this full Bayesian matrix factorization approach (BPMF in the following) are represented in Figure 3.12.

Exact evaluation of the predictive distribution in Equation 3.20 is analytically intractable, due to the complexity of the posterior term $P(\mathbf{P}_u, \mathbf{Q}_i|\mathcal{X}, \Theta_U, \Theta_I)$. Hence, the need for approximate inference, which is approached in [168] by adopting Gibbs sampling. In particular, the predictive distribution is approximated as

$$P(r|u, i, \mathcal{X}, \Theta_0) \approx \frac{1}{T} \sum_{t=1}^{T} P(r|\mathbf{P}^t, \mathbf{Q}^t, \sigma),$$

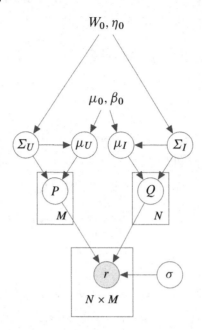

Figure 3.12: Graphical model for BPFM.

where the matrices $\{\mathbf{P}^t, \mathbf{Q}^t\}_{t=1,\ldots,T}$ are computed by using a Markov chain whose stationary distribution is the posterior of the model parameters $\{\mathbf{P}, \mathbf{Q}, \Theta_U, \Theta_V\}$. The Gibbs sampling procedure is devised by the following learning steps:

1. Initialize $\{\mathbf{P}^1, \mathbf{Q}^1\}$ randomly.

2. For each $t \in \{1, \ldots, T\}$:

   (a) Sample $\Theta_U^t \sim P(\Theta_U | \mathbf{P}^t, \Theta_0)$ and $\Theta_I^t \sim P(\Theta_I | \mathbf{Q}^t, \Theta_0)$;

   (b) For each $u \in \mathcal{U}$ sample $\mathbf{P}_u^{t+1} \sim P(\mathbf{P}_u | \mathcal{X}, \mathbf{Q}^t, \Theta_U^t)$;

   (c) For each $i \in \mathcal{I}$ sample $\mathbf{Q}_i^{t+1} \sim P(\mathbf{Q}_i | \mathcal{X}, \mathbf{P}^{t+1}, \Theta_I^t)$.

Within this sampling scheme, the adoption of conjugate priors eases the sampling steps. In particular, $P(\mathbf{P}_u | \mathcal{X}, \mathbf{Q}^t, \Theta_U^t)$ and $P(\mathbf{Q}_i | \mathcal{X}, \mathbf{P}^{t+1}, \Theta_I^t)$ are still multivariate Gaussian, whereas $P(\Theta_U | \mathbf{P}^t, \Theta_0)$ and $P(\Theta_I | \mathbf{Q}^t, \Theta_0)$ are Gaussian-Wishart.

3.5 SUMMARY

Techniques based on Bayesian modeling are better suited when the underlying data is characterized by high sparsity (like in the case of recommender systems), as it allows a better control of the priors that govern the model and it prevents overfitting. Moreover, the introduction of informative priors allows us to exploit prior knowledge, as it easily can be encoded in the hyperparameters of the prior distributions. Bayesian methods have been successfully applied for modeling preference data and in the following table, we provide a summary of the main features of the approaches discussed in this chapter. All of them employ a soft membership policy, which allows us to allocate different latent factors to different preference observations. LDA focuses on the modeling of implicit data, hence, does not support the task of rating prediction. Also, it is worth noticing that, although co-clustering and matrix factorization approaches employ latent factor modeling on both the dimensions of the preference matrix, MF techniques model, for each data observation, user and item factors independently. Table 3.4 summarizes the basic properties of the proposed models according to the various dimensions.

Table 3.4: Summary

		LDA	URP	BUCM	Bi-LDA	RBC	BH-Free	BH-Forced	PPMF	BPMF
PREDICTION MODE	Free			✓			✓			
	Forced		✓		✓	✓		✓	✓	✓
FACTORIZATION DIMENSION	User-based	✓	✓	✓	✓	✓				
	Item-based				✓	✓				
	Joint								✓	✓
	Hierarchical						✓	✓		
PREDICTION MODEL	Discrete	✓	✓	✓	✓		✓	✓		
	Continuous		✓			✓			✓	✓
FACTORIZATION SCOPE	Hard									
	Soft	✓	✓	✓	✓	✓	✓	✓	✓	✓
PREFERENCE MODELING	Implicit feedback	✓								
	Explicit (Multinomial)		✓	✓	✓		✓	✓		
	Explicit (Gaussian)					✓			✓	✓

The practical advantages in employing Bayesian techniques over non-Bayesian approaches in prediction accuracy are highlighted in Chapter 4. There, we extensively compare the performances in rating prediction and recommendation and discuss the capabilities of these methods.

The wide and successful application of Bayesian methods is mainly motivated by the development of effective approximate inference algorithms (such as variational inference and expectation propagation) and sampling techniques (such as Markov chain Monte Carlo and collapsed Gibbs sampling). There is an ongoing debate over the choice of approximate inference techniques over sampling methods. Variational inference techniques aim at providing deterministic approximations for the true posteriors; they are usually harder to derive than sampling-based techniques and there is not a clear recipe to derive such approximations. Their main drawback is that the loose approximation of posteriors tends to produce inaccurate results. Sampling-based methods aim at approximating, by Monte Carlo simulations, the predictive distribution of the considered model; the distributions that specify the model are iteratively updated by sampling from a Markov chain, whose stationary distribution is an approximation of the posterior over the model parameters and hyperparameters. This approximation asymptotically converges to the true posterior, but it is hard to determine the convergence of the Markov chain.

The scalability of inference procedures is a main line of research that follows two main directions. The first one focuses on directly speeding up learning procedures by accelerating the computation of the exact solution on a single machine. In this context, effective speeding-up techniques have been developed for the LDA collapsed Gibbs sampling. The main idea is to reduce the time taken for sampling a topic for each observation. This sampling operation typically involves the computation of the posterior probability of each topic given the current observation and the remaining parameters (see Equation 3.8); hence, it is linear in K, the number of topics. However, for any particular observation, the posterior distribution is frequently skewed, i.e., the probability mass is concentrated on a small fraction of topics. It is possible [156] to exploit this property by iterating the list of topics sorted by decreasing popularity, and computing iteratively the probability that the sampled value, for a given observation, does not depend on the remaining topics. This process allows us to relate each sample to a limited subset of topics, thus substantially speeding up the learning phase without introducing approximations. It is shown in [156] that this approach is eight times faster than standard collapsed Gibbs sampling for LDA.

The other line of research involves the development of parallel implementations of learning algorithms in a multiprocessor architecture and/or in a distributed setting. The expectation-maximization algorithm (as well as variational EM) easily can be parallelized by considering that responsibilities (or variational parameters) can be computed independently for each document (or user). In a multiprocessor architecture [140], the E-step can be executed by using parallel threads, and a join thread can aggregate the sufficient statistics produced by each thread and update the model parameters in the M-step.

Similarly, in a master-slave distributed environment, it is possible to partition the collection into blocks (subsets of documents) that are processed independently; the synchronization step is performed by a master node, which receives all the sufficient statistics from other nodes, updates the model, and writes it on the disk. The computation performed by the master node, which is the main bottleneck in this approach, can be further parallelized. For example, in [206] the authors propose an implementation of variational inference for LDA based on the *Map-Reduce* framework. The computation is distributed across a set of *mapper* and *reducer* nodes: the former process small subsets of data performing the variational E-step, while the latter aggregate the partial results produced by mappers and update the model. The computational burden of the M-step is addressed by the fact that each reducer receives and processes the statistics corresponding to a single topic. A similar idea can be used to speed-up the learning phase of variational Bayesian PMF[192].

The task of distributing and parallelizing learning algorithms based on Gibbs sampling has an intrinsic difficulty that comes from the sequential nature of the sampling chain. In fact, the asymptotical convergence to the stationary distribution requires the sequential updates of topic assignments: the probability of sampling a topic for one observation depends on the topics assigned to all the previous observations in the sampling chain. In [143], authors study a parallel Gibbs sampling algorithm for LDA in which this dependency is relaxed. Each processor receives the current status of global counts, performs Gibbs sampling on the assigned subset of data, and updates its local table with topic assignments and counters. At the end of this computation, a synchronization thread gathers local statistics from samplers and updates the global status of counts by enforcing consistency. Despite having no formal convergence guarantees, due to the fact that we are no longer sampling from the true posterior, this parallel (and approximate) implementation works well in practice.

We conclude this chapter by mentioning that several implementations of scalable learning algorithms for Bayesian probabilistic latent factor models are freely available. Among them we recall the following.

- **Mahout**[2] provides an implementation of collapsed variational Bayes for LDA [191] on Apache Hadoop using the map/reduce paradigm.

- **Vowpal wabbit**[3] provides an online implementation of LDA based on [80]: training data is divided in mini-batches, for each document in the current batch the algorithm computes the E-step and updates the approximated topic distributions.

- **GraphLab**[4] provides a parallel implementation of the collapsed Gibbs sampling for LDA based on the work described in [6], which allows effective synchronization and storage of the state of latent variables and is able to fit streaming data.

- **Fast And Scalable Topic-Modeling Toolbox**[5] implements several parallel and distributed algorithms for LDA, based on both fast collapsed variational Bayesian inference [13] and collapsed Gibbs sampling [143, 156].

---

[2]https://mahout.apache.org/
[3]https://github.com/JohnLangford/vowpal_wabbit/wiki
[4]http://graphlab.org/projects/index.html
[5]http://www.ics.uci.edu/~asuncion/software/fast.htm

CHAPTER 4

Exploiting Probabilistic Models

As discussed in Chapter 2, probabilistic approaches can be classified according to two alternative ways of modeling preference observations:

- **Forced prediction**: This model provides the estimate of $P(r|u, i)$, which represents the conditional probability that user u assigns a rating value r given the item i;

- **Free prediction**: The process of item selection is included in this model, which is typically based on the estimate of $P(r, i|u)$. In this case, we are interested in predicting both the selection of the item and the corresponding preference of the user. By applying the chain rule, $P(r, i|u)$ can be factorized as $P(r|i, u)P(i|u)$; this factorization still includes a component of forced prediction, which, however, is weighted by the probability of selecting the item and thus allows a more precise estimate of user's preferences.

The exploitation of a probabilistic model in a recommendation scenario relies on the analysis of the distribution over preference values expressed by the above probabilities. Given this distribution, there are several methods for computing, for each user, a personalized ranking over items. In the following, we will discuss possible instantiations of ranking functions that exploit the flexibility of the approaches analysed in the previous chapters.

Most CF techniques have focused on the development of accurate techniques for rating prediction. The recommendation problem has been interpreted as a missing-value prediction problem [174], in which, given an active user, the system is asked to predict her preferences on a set of items. However, [47, 126] have shown that the focus on prediction accuracy does not necessarily help in devising good recommender systems.

As the number of available customers increases, it is always more difficult to understand, profile, and segment their behaviors and a similar consideration holds for the catalog of products. Under this perspective, CF models should be considered in a broader sense, for their capability to deeply understand the hidden relationships among users and the products they like. For example, the high-dimensional preference matrix can be partitioned to detect user communities and item categories; the analysis of the relationships between these two levels can provide a faithful, yet compact, description of the data that can be exploited for better decision making.

4.1 PROBABILISTIC MODELING AND DECISION THEORY

The modeling of data observations with probability distribution that employ random variables provides us with a mathematical framework for quantifying uncertainty. For instance, in a recommendation scenario, the joint distribution $P(u, i, r)$ summarizes the degree of uncertainty corresponding to the observation $\langle u, i, r \rangle$. From an application perspective, we face the problem of performing choices under uncertainty. That is, we are interested in exploiting this measure of uncertainty to make practical decisions: given a probability distribution over ratings for each item i not yet purchased by the current user, how do we select the next items to be recommended? This can be seen as a two-step process: in the first step, we make a specific prediction over the random variable of interest (e.g., the predicted rating value for the pair $\langle u, i \rangle$), while, in the second step, we perform decisions based on such predictions.

The process can be formalized according to [138, Section 5.7], as follows. We are given a set of observations $\mathcal{X}$, and each observation $\mathbf{x} \in \mathcal{X}$ can be associated with an unknown state $y \in \mathcal{Y}$. Decisions can be encoded as possible actions $a \in \mathcal{A}$, and a loss function $L(y, a)$ measures the compatibility of action a relative to the hidden state y. Some example are, e.g., the 0/1 loss $L(y,) = \mathbb{1}[\![y \neq a]\!]$, the square loss $L(y, a) = (y - a)^2$, or the logarithm loss $L(y, a) = -\log a$. For our purposes, we encode actions based on a specific parametric probabilistic model. Each model $\mathcal{M}$ encodes a possible decision $\delta(\mathbf{x}, \mathcal{M})$ for a given observation $\mathbf{x}$. This allows us to define the notion of *risk* as

$$\mathcal{R}(\mathcal{M}) = \sum_{\mathbf{x} \in \mathcal{X}} L(y, \delta(\mathbf{x}, \mathcal{M})),$$

or, in more general terms,

$$\tilde{\mathcal{R}}(\mathcal{M}) = \sum_{\mathbf{x} \in \mathcal{X}, y \in \mathcal{Y}} P(y, \mathbf{x}) L(y, \delta(\mathbf{x}, \mathcal{M})).$$

We have mentioned, in the beginning of this chapter, that recommendation has been traditionally interpreted as missing rating prediction. In this context, $\mathbf{x}$ represents an observation (u, i), and an action $\delta((u, i), \mathcal{M})$ can correspond to optimal rating assignment for the observation, based on the underlying model. In probabilistic terms, this can be encoded as $\delta((u, i), \mathcal{M}) = \hat{r}_i^u$, and $\hat{r}_i^u$ where

$$\hat{r}_i^u = \underset{r}{\text{argmax}} \, P(r|u, i, \mathcal{M}), \tag{4.1}$$

or, alternatively,

$$\hat{r}_i^u = E[R|u, i; \Theta] = \int r P(r|u, i, \mathcal{M}) \, dr. \tag{4.2}$$

It can be shown [138] that, for a fixed model $\mathcal{M}$, Equation 4.1 is optimal for the 0/1 loss, and Equation 4.2 is optimal for the square loss.

The risk function also can be used for parameter estimation, when actions represent an estimation of some values on the data, and the loss function measures the difference between estimated and true values. In such cases, there are some interesting connections between the risk function and the likelihood functions described in Chapters 2 and 3. For example, if we focus on preference observations and consider the PMF approach discussed in Section 2.3.2, we can observe that the log likelihood corresponds to the empirical risk under the square loss, with $\delta(\langle u, i, r \rangle, \mathcal{M})$ defined as in Equation 4.2. More in general, we can adopt $\delta(\langle u, i, r \rangle, \mathcal{M}) = P(r|u, i, \mathcal{M})$. Then, under a logarithm loss, we obtain

$$\mathcal{R}(\mathcal{M}) = \sum_{\mathbf{x} \in \mathcal{X}} L(y, \delta(\mathbf{x}, \mathcal{M}))$$

$$= - \sum_{\langle u,i,r \rangle} \log P(r|u, i, \mathcal{M})$$

$$= - \log P(\mathcal{X}|\mathcal{M}).$$

In this case, maximizing the likelihood corresponds to minimizing the empirical risk. Besides the relationships between likelihood and empirical risk, it is, in general, possible to reformulate the learning problem as the minimization of application-specific losses: for example, in terms of losses encoding RMSE [131] or ranking [130, 159].

We can summarize the above discussion in the following observations.

- On one side, we assume that models are given and learned in some manner, and we can exploit the relationship between models and actions to compare models according to a given criterion that is encoded by a loss function.

- On the other side, given a model, we can directly encode the decision process as dependent on the model parameters. Then, we can find the optimal parameters that minimize the risk associated with the decisions made under that model.

Given an estimate of the probability distributions that minimize some prediction loss, we next turn our attention to the problem of making final decisions. Although, in theory, the direct inference of parameters that minimizes the considered risk function could be considered as the best choice from an application-specific perspective, the choice of decoupling inference from decisions is more suitable in several scenarios [31, Section 1.5]:

- **Computational tractability**. Even if we described above some situations in which minimizing the risk corresponds to maximizing a formulation of the log likelihood, there are settings, e.g., if we specify a loss function directly on the recommendation list, where optimal decisions cannot be analytically derived or approximated.

- **Flexibility**. Generally, the estimate of the posterior distribution of the parameters of the model-given observations provides us several ways of performing decisions; by addressing inference and decisions at the same time, we focus exclusively on one way of exploiting the model and this could essentially result in a loss of information.

- **Revision and weighting of the loss function**. If we consider a situation where the loss depends on financial expectations, and the latter are subjected to revision from time to time, decisions easily can be adapted by tuning the decision surfaces rather than the set of parameters. More generally, the loss can depend on a weighting function that may change with time. Thus, separating decisions from inference allows us to tune the first ones without having to infer the model again.

- **Multiple objective functions**. We can consider the case where the risk function can break into a number of components, each of which represents a separate subspace that can be tackled separately. Rather than combine all of this heterogeneous information, it may be more effective to work on the different pieces separately, by modeling the likelihood of each subspace and then by combining them by exploiting ad-hoc decisions for each subspace.

- **Model comparison**. Given two models $\mathcal{M}_1$ and $\mathcal{M}_2$, comparing such models may depend on the underlying loss function. In particular, there may be situations where $\mathcal{M}_1$ outperforms $\mathcal{M}_2$ in predicting the rating, while the latter provides a more accurate recommendation list. We shall see that, in such situations, relying on a single optimization strategy can yield inappropriate decisions.

In the following, we evaluate the approaches presented in Chapters 2 and 3 in different settings, such as prediction and recommendation accuracy, which reflect different risk functions. More generally, probabilistic techniques offer some renowned advantages: notably, they can be tuned to optimize a variety of risk functions; moreover, optimizing the likelihood allows us to model a distribution over rating values, which can be used to determine the confidence of the model in providing a recommendation and to implement a *rejection policy* where no decision can be made with enough confidence. Finally, they allow the possibility to include prior knowledge in the generative process, thus allowing a more effective modeling of the underlying data distribution.

Specifically, we are interested in detecting which models and assumptions better support specific losses over others. Investigating such aspects allows us to define a key for better recommendation according to the underlying scenario we are interested in modeling.

4.1.1 MINIMIZING THE PREDICTION ERROR

We base our analysis on the *Movielens1M*[1] benchmark dataset, which was collected in the context of "The GroupLens Research Project" developed within the Department of Computer Science and Engineering at the University of Minnesota. This dataset contains 1,000,209 anonymous ratings on approximately 3,700 movies made by 6,040 users and it is widely used as a benchmark for evaluating recommendations and CF techniques. This collection contains tuples $\langle u, i, r, t \rangle$, where u and i are, respectively, user and item identifiers, t is a timestamp, $r \in \{1, \ldots, 5\}$ is the rating value, and each user has at least 20 ratings. Side information is also available, both for users (gender, age, occupation, Zip-code) and movies (title, genre). Table 4.1 contains a summary of the main properties of the data.

Figure 4.1 shows the cumulative distribution relative to the number of ratings per item and user. In both cases, we can observe a extremely skewed distributions: a small percentage of

---

[1]http://www.grouplens.org/datasets/movielens

Table 4.1: Movielens1M—statistics

#users	6,040
#items	3,706
#preferences	1,000,209
Avg #ratings user	166
Avg #ratings item	270
Min #ratings user	20
Max #ratings user	2,314
Min #ratings item	1
Max #ratings item	3,428

items/users is associated to a large majority of ratings. In particular, as shown in Figure 4.1(b), 10% of the items (resp. users) involve 40% of the ratings. Finally, Figure 4.2 shows the distribution of ratings and highlights a clear bias toward 4 stars, and a general tendency to provide positive evaluations.

In the following, we shall analyze the behavior of the algorithms described in the previous chapters with regard to RMSE, which corresponds to computing the average risk relative to the squared loss. In the following experiments, we perform a chronological split of the rating matrix, by retaining the first 80% of the triplets as training data for learning the models, and using the remaining 20% as test data for measuring the performance.

Mixture Models and Rating Prediction.   In a first set of experiments, we compare some basic approaches based on latent factor modeling. These include: the *Mixture Model* (MMM); *User Community Model* (both Gaussian and multinomial; UCMG/UCMM); *Probabilistic Latent Semantic Analysis* (GPLSA/PLSA); and *Probabilistic Matrix Factorization* (PMF).[2] For all these methods, we study the performances in RSME for increasing numbers of latent factors. In particular, we range the latter within the values $\{2^K\}_{K=1,\ldots,7}$. Figure 4.3 reports both performances in prediction accuracy and running times.

We can see a substantial difference between Gaussian and multinomial approaches. In particular, the latter tend to produce a higher error rate, especially with a high number of factors. On the contrary, Gaussian models tend to outperform all the remaining approaches. The explanation is twofold. First, the specification of the likelihood in the Gaussian models tends to include a component that relates to the RMSE directly. Hence, optimizing the likelihood corresponds to optimizing the RMSE as well. Secondly, RMSE tends to penalize large errors in predictions: that is, predicting four stars where the actual value is one has a higher impact on the RMSE than predicting two stars. This effect is not reflected in the log likelihood corresponding to multinomial models, while it is fully considered by Gaussian models.

[2]The code for PMF is kindly provided by Russian Salakhutdinov at http://www.utstat.toronto.edu/~rsalakhu/code.html.

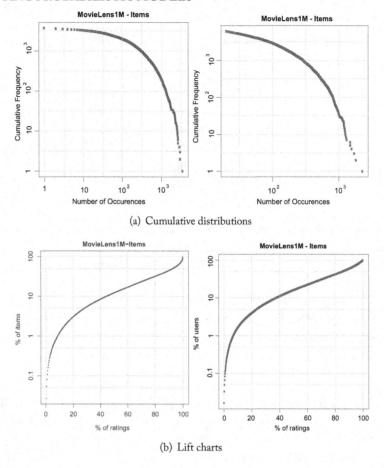

(a) Cumulative distributions

(b) Lift charts

Figure 4.1: Distribution of evaluations in MovieLens.

In an attempt to carefully analyze the rating prediction pitfalls, in Figure 4.3(b), we plot the contribution to the RMSE in each single evaluation in $\mathcal{V}$ by the probabilistic techniques under consideration and considering Item-Avg acts as baseline. Predictions exhibit a good accuracy for values 3–4, and their performance is weak on border values, namely 1, 2, and 5. This is mainly due to the nature of RMSE, which penalizes larger errors.

The learning time is shown in Figure 4.3(c). All methods scale linearly in the number of topics, and the UCM approaches exhibit the largest overhead.

We now turn our attention to the co-clustering approaches, where latent factors components exhibit a more complex structure. Notably, complex patterns can be better detected by means of co-clustering approaches, as the latter aim at partitioning data into homogeneous blocks, enforcing a simultaneous clustering on both dimensions of the preference data. This highlights the mutual relationships between users and items. In this respect, co-clustering approaches should

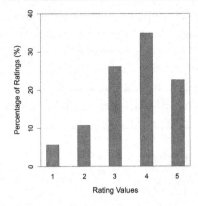

Figure 4.2: Distribution of ratings in MovieLens.

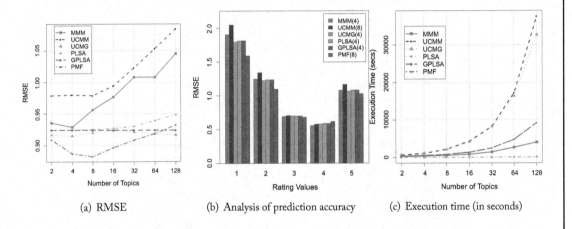

(a) RMSE   (b) Analysis of prediction accuracy   (c) Execution time (in seconds)

Figure 4.3: Performance on MovieLens.

be considered in a broader sense, for their capability to deeply understand the hidden relationships among users and products they like. Examples in this respect are user communities, item categories, and preference patterns within such groups. Besides their contribution to the minimization of the prediction error, these relationships are especially important, as they can provide a faithful yet compact description of the data, which can be exploited for better decision making.

Surprisingly, co-clustering approaches seem able to provide accurate predictions as well, and they represent an upgrade of the basic latent factor methods. Figure 4.4 shows the contour plot of both the BMM and the FMM for increasing values of user and item factors. The performance of BMM clearly improves that of its uni-dimensional counterpart MMM, and is in line with the performance of GPLSA and UCMG. On the other side, FMM provides extremely good results for an increasing number of topics. This is a behavior we observed before, with GPLSA and UCMG as well, in Figure 4.5. Apparently, these methods give better results with more topics, which, in a sense, is unsatisfying and intuitively not natural. We shall comment on this peculiarity further in this chapter.

The computational overhead due to the complexity of the learning procedure is apparent in Figures 4.4(c) and 4.4(d). In particular, the timings for the FMM model are extremely high. This is somehow surprising, since FMM appears to be an upgrade of the PLSA model.

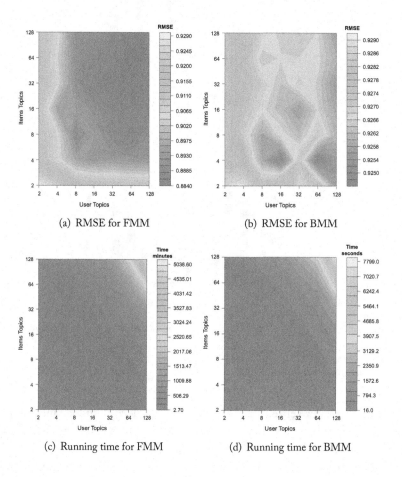

(a) RMSE for FMM   (b) RMSE for BMM

(c) Running time for FMM   (d) Running time for BMM

Figure 4.4: Performance of co-clustering approaches.

Bayesian Modeling and Rating Prediction.   In this section we analyze the Bayesian formulations discussed in Chapter 3. While traditional factor modeling relies on maximum likelihood estimation for model inference, the Bayesian formulations are better suited to the sparsity of the rating matrix and less susceptible to overfitting. In this respect, we expect that the best performances can be obtained by exploiting these models.

Once again, the methods based on matrix factorization outperform the other methods. As already explained, the likelihood of these models has a direct correspondence with the squared loss minimization. In addition, Bayesian marginalization plays a major role in improving the results. This is evident if we compare the performance of PMF with those of BPMF[3] and PPMF,[4] which are Bayesian upgrades of the former.

By contrast, neither URP nor BUCM exhibit competitive performances. As for the BUCM, there are two aspects to consider. First, it provides a multinomial model for the ratings, which we have already seen are weaker than Gaussian models. Second, the focus of this model is on free prediction, and, hence, the resulting likelihood does not directly optimize the probability $p(r|u, i)$, which is, conversely, necessary for accurate predictions. Also, the performance of URP is somehow disappointing, if we consider that the latter is aimed at optimizing the likelihood of a forced prediction $P(r|u, i)$ that is still modeled as a Gaussian.

As for what the scalability is concerned, it seems that the variational approximation in the PPMF introduces a significant overhead over the other methods, as we can see from Figure 4.3(c). As for the URP, the overhead essentially is due to the fact that the sampling procedure does not reach a fast convergence, and it is stopped after 3,000 iterations.

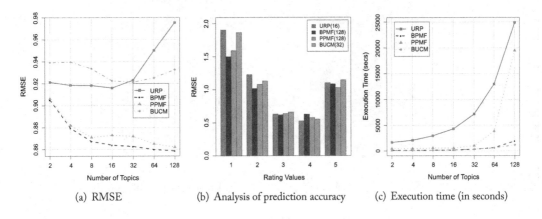

(a) RMSE   (b) Analysis of prediction accuracy   (c) Execution time (in seconds)

Figure 4.5: Performance on MovieLens.

Finally, we present the performance of the Bayesian hierarchical methods in Figure 4.6. Since these models were introduced to accommodate hierarchical modeling and rating prediction, the predictive accuracy exhibited by both models over unobserved ratings is comparable and, in some cases, even better than the other probabilistic approaches illustrated so far. Interestingly, the general trend shown by the other co-clustering approaches is confirmed. As for the BH-free

---

[3]The code is again provided by Russian Salakhutdinov in `http://www.utstat.toronto.edu/~rsalakhu/code.html`.
[4]Available at `http://www-users.cs.umn.edu/~shan/ppmf_code.html`.

approach, it is worth noticing that it overcomes the BUCM approach: since both methods model the rating in a similar manner, this result shows that the hierarchical structure provides substantial information for boosting the accuracy of prediction. Again, these methods exhibit a significant computational overhead over "simpler" methods, as witnessed by the Figures 4.6(c) and 4.6(d).

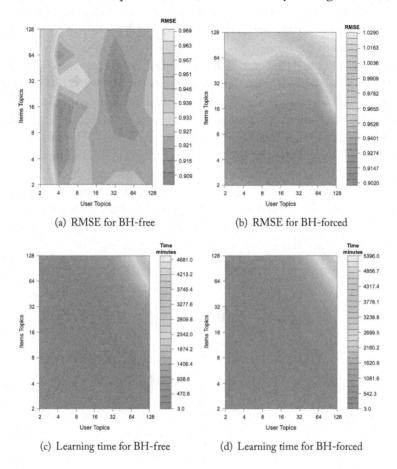

(a) RMSE for BH-free　　　　　　　(b) RMSE for BH-forced

(c) Learning time for BH-free　　　　(d) Learning time for BH-forced

Figure 4.6: Bayesian hierarchical co-clustering and RMSE.

We conclude this section by summarizing the results obtained in the above experiments: Table 4.2 reports, for each method discussed in the section, the best RMSE achieved and the required number of latent factors exploited.

4.1.2 RECOMMENDATION ACCURACY

The testing methodology based on the minimization of statistical error metrics such as RMSE has important limitations, mainly because the assumption *"improvements in RMSE would reflect into*

Table 4.2: Summary results for RMSE

Model	Number of Topics	RMSE
MMM	4	0.9287
UCMM	8	0.9788
UCMG	4	0.9148
PLSA	4	0.9242
GPLSA	4	0.9241
URP	16	0.9159
PMF	8	0.8821
BPMF	128	0.8585
PPMF	128	0.8621
BUCM	32	0.9212
FMM	64×128	0.8843
BMM	16×4	0.9247
BH-Free	32×32	0.9085
BH-Forced	4×8	0.9025

better recommendations" does not necessarily hold. In [47], Cremonesi et al. review and compare the most common CF approaches to recommendation in term of the accuracy of the recommendation lists, rather than on the rating prediction accuracy based on RMSE. Their analysis shows that cutting-edge approaches, characterized by low RMSE values, achieve performances comparable to naive techniques, whereas simpler approaches, such as the pure SVD, consistently outperform the other techniques. In an attempt to find an explanation, the authors impute the contrasting behavior with a

> *"limitation of RMSE testing, which concentrates only on the ratings that the user provided to the system, [and consequently] misses much of the reality, where all items should count, not only those actually rated by the user in the past." [47]*

Pure SVD rebuilds the whole preference matrix in terms of latent factors, by considering both observed and unobserved preference values. This approach better identifies the hidden relationships between both users and items, which in turn results in a better estimation of the item selection.

In an attempt to better analyze this aspect, we can turn our attention again on the distinction between forced and free prediction and see how these alternative modeling choices can impact the ranking p_i^u exploited in the evaluation protocols discussed in Chapter 1, with particular reference to the Algorithm 1 and 2.

Predicted Rating.   When explicit preferences are available, the most intuitive way to provide item ranking in the recommendation process relies on the analysis of the distribution over preference values $P(r|u, i)$. Given this distribution, there are several methods for computing the ranking for each pair $\langle u, i \rangle$; here again, we resort to the expected rating $p_i^u = E[R|u, i]$ defined in Equation 4.2. The idea is that a recommendation list should contain only items that a user will like. Hence, if two items achieve a different expected rating, by comparing the latter we can provide a way to express preferences in the recommendation list. Alternative measures can consider whether a given item will obtain a high preference, i.e.,

$$p_i^u = P(r \geq t | u, i) = \int_t^{+\infty} P(r|u, i) \, dr,$$

where t represent a rating threshold: for example, by setting $t = \bar{r}_T$, we can rank items by the probability that they'll get a rating higher than the mean rating.

Following [47], we consider *Item-Avg* and *Top-Pop* as baselines and compare some selected approaches with respect to *Pure-SVD* matrix approximation with 50 factors. We then consider again the PMF, BPMF, PLSA, URP, and BH-forced. All these methods are characterized by their capability of computing the above mentioned ranking functions. By exploiting the respective ranking, we then compute the recommendation accuracy by considering a sample size $D = 1,000$ and an increasing size of the recommendation list. The behavior of these algorithms is shown in Figure 4.7.

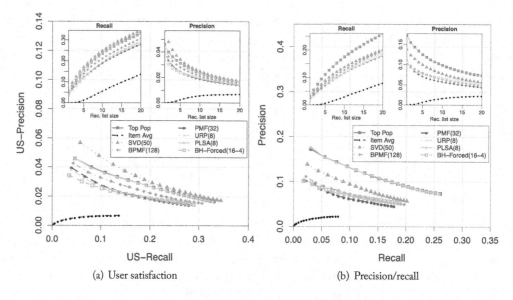

(a) User satisfaction   (b) Precision/recall

Figure 4.7: Recommendation accuracy, exploiting the expected rating.

This brief analysis confirms that there is no monotonic relationship between RMSE and recommendation accuracy. Considering user satisfaction, almost all the probabilistic approaches fall between the two baselines. Pure SVD significantly outperforms the best probabilistic performers, namely BPMF and PLSA. The trend for probabilistic approaches does not change considering recall and precision, but in this case not even the pure-SVD is able to outperform Top-Pop, which exhibits a consistent gain over all the considered competitors. Also, it's surprising that the PLSA model, particularly weak in RMSE, outperforms the matrix factorization approaches.

In general, by comparing the approaches to the baselines, we can conclude that ranking by the expected value does not seem an adequate tool for providing accurate recommendation lists, nor does any variant of this approach.

In an attempt to carefully analyze the rating prediction pitfalls, we can consider again Figure 4.3(b). The penalization of border values clearly supports the thesis that low RMSE does not necessarily induce good accuracy, as the latter is mainly influenced by the items in class 5 (where the approaches are more prone to fail). It is clear that a better tuning of the ranking function should take this component into account. Also, by looking at the distribution of the rating values, we can see that the dataset is biased toward the mean values, and more generally the low rating values represent a lower percentage. This explains the tendency of the expected value to flatten toward a mean value (and hence to fail in providing an accurate prediction). On the other hand, low values are rarer in the dataset. That is, people are inclined to provide only good ratings, thus making the ranking based on preference values less reliable.

Combining Item Selection and Relevance.   The rank p_i^u necessary for a recommendation list, as described in Chapter 1, does not necessarily depend on the prediction of an explicit rating value $\hat{r}_i^u$. In particular, for approaches based on the modeling of implicit preferences, the rank of each item i, with regards to the user u, can be computed as the mixture:

$$p_i^u \triangleq P(i|u) = \sum_{k=1}^{K} P(k|u)P(i|k),$$ (4.3)

where $P(i|k)$ is the probability of selecting i within the topic k.

When ratings are available, we can still boost the ranking of models that directly support free selection, by forcing the selection process to concentrate on relevant items. For example, we can upgrade either the expected rating,

$$p_i^u \triangleq \int r \cdot p(r, i|u)\, dr = P(i|u)E[r|u, i],$$

or the likelihood for a high preference,

$$p_i^u \triangleq P(i, r > \bar{r}_{\mathbf{T}}|u) = P(i|u)P(r > \bar{r}_{\mathbf{T}}|u, i).$$ (4.4)

In both cases, the rank of an item considers the probability that an item is selected, and, hence, higher ranks are provided to items more likely to be selected.

Including the item selection component in the ranking has an apparent counter-intuitive effect: popular average-rated items are likely to get better ranking than niche, high-rated items. Yet, explicit modeling of item selection plays a crucial role with accurate recommendation lists.

This effect has been studied in details in [17, 19, 20] and we summarize the main results in Figure 4.8.

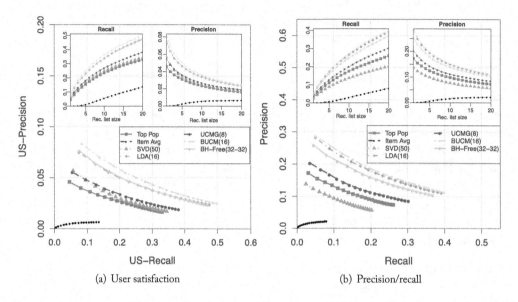

(a) User satisfaction   (b) Precision/recall

Figure 4.8: Recommendation accuracy achieved by probabilistic approaches considering item selection and relevance as ranking functions.

Figure 4.8 compares the recommendation accuracy achieved by probabilistic models that employ item-selection (LDA, Equation 4.3) and item-selection & relevance (BUCM, UCM, and BH-free, Equation 4.4). This evaluation confirms the importance of the selection component in the recommendation process. All the approaches significantly outperform the pure SVD, and the best result is obtained by the approach combining the item selection and relevance in a Bayesian setting (Bayesian UCM). Notably, a quite simple model like LDA obtains a consistent gain with regard to both the SVD and the Top-Pop. Compared to the results in Figure 4.7, these results show the apparent mismatch between two opposed situations: the explicit modeling of item selection boosts the accuracy of recommendation lists, although it seems to negatively impact on prediction accuracy. The latter is mainly due to the underlying optimization strategy: optimizing the likelihood in these models does not have a direct correspondence with the optimization of the RMSE, like it happens, e.g., in the approaches based on matrix factorization. As a consequence, these models are more general and they provide a better response in recommendation accuracy.

4.2   BEYOND PREDICTION ACCURACY

As discussed in Chapter 1, the accuracy of a recommender system is just one of the relevant dimensions in evaluating its effectiveness: concepts like novelty, serendipity, diversity, and coverage are very important for the recommendation quality improvement and they are still a matter of study. As users may exhibit interest in different topics, it is important to provide them with recommendation lists that cover their topics of interests, as well as to avoid monotonous recommendations [207], by maximizing the diversity of the lists provided in each session.

Here, we focus on the application of probabilistic technique to the aforementioned dimensions of analysis. As we show in Section 4.2.1, the latent variable modeling, exploited within a probabilistic framework, offers some important, and easily interpretable, insights into the users' purchase and preference patterns. In Section 4.1.2 we study how the data analysis can be extended to the case of co-clustering, and, finally, in Section 4.2.3 we discuss the application of probabilistic topic models for the generation of diverse recommendations.

4.2.1   DATA ANALYSIS WITH TOPIC MODELS

Identifying groups of similar-minded users is a powerful tool for developing targeted marketing campaigns for communities of users who exhibit the same preference patterns. One of the main advantages of probabilistic latent factor models is to provide a compact representation of the data, by highlighting hidden relationships between the modeled entities, which, in our case, are users and items. In this respect, models can be used to summarize preference patterns in terms of latent topics, which, in turn, can be exploited to detect communities of like-minded users or similar items.

Towards this goal, we can apply the methodology proposed in [100], which focuses on the analysis of implicit feedback (web sessions), by applying PLSA to (i) understand users' preference and interests, (ii) infer the underlying task of a web-browsing session, and (iii) discover hidden semantic relationships between users and resources (web pages, products, etc.). Assume we have a dataset of implicit co-occurrence pairs $\langle u, i \rangle$, where each pair encodes the fact that the corresponding user u purchased/viewed/consumed resource i. Preference patterns easily can be analyzed if the underlying probabilistic model enables the estimation of the components $P(k|u)$ (representing the interest of the user in the k-th topic), $P(u|k)$ (representing the probability of observing u relative to topic k), the dual $P(i|k)$ (representing the probability of observing u relative to topic k), and a prior probability $P(k)$ that a user is associated with topic k, which generally can be obtained by marginalizing over users:

$$P(k) = \sum_u P(k|u)P(u).$$

Labeling Topics.   Some specific topic models, such as PLSA or LDA, associate latent factors with a specific predictive distribution over items, modeling an abstract preference pattern where some items are most probable given a topic. However, these models do not associate any automatic *a priori* meaning with the states of the latent factor variables. One of the most important challenges in topic analysis is to associate each topic with a set of meaningful labels that can be used to interpret the underlying semantic concept. The task of topic labeling is generally manual:

the analyst subjectively selects a set of labels that match the category (theme, genre) of the more likely items for each considered topic.

In the context of CF data, this task can be facilitated by exploiting the predictive distribution in order to provide a selection of *characteristic* items. Intuitively, a characteristic item for a given topic k is an item that exhibits a high probability of being observed, given the considered topic, and a low probability, given a different topic. According to [100], we can rank items according to the following score:

$$score(i, k) = P(i|k)P(k|i),$$

where

$$P(k|i) = \frac{P(i|k)P(k)}{\sum_{k'} P(i|k')P(k')}.$$

Then, characteristic items for the topic k are those items associated with the highest score. The top-10 characteristic items for Movielens data are provided in Table 4.3.

The task of topic labeling can be also be cast as a discrete optimization problem [129]. Let $\mathbf{L}$ be a list of items and assume that we can compute the probability $P(i|\mathbf{L})$. Then, $\mathbf{L}$ is a good representative set for k if the latter probability is a good approximation of $P(i|k)$ in terms of Kullback-Leibler (KL) divergence:

$$KL(k||\mathbf{L}) = -\sum_i P(i|k)\frac{P(i|k)}{P(i|\mathbf{L})}.$$

Now, $P(i|\mathbf{L})$ can be defined by resorting to the set $\mathcal{S}(\mathbf{L}) = \{u \in \mathcal{U} | \mathcal{I}_{\mathbf{R}}(u) \cap \mathbf{L} \neq \emptyset\}$ of all users that chose items in $\mathbf{L}$ before. Then

$$P(i|\mathbf{L}) \triangleq \frac{1}{|\mathcal{S}(\mathbf{L})|} \sum_{u \in \mathcal{S}(\mathbf{L})} \sum_k P(i|k)P(k|u),$$

where the latter definition encodes the intuition that an item is probable in $\mathbf{L}$ if it is likely in all users that adopted items in $\mathbf{L}$ before.

Identify Communities of Like-Minded Users.   The projection of each user profile into the latent topic space can be used to detect communities of users who exhibit the same preference patterns. Given a set of K topics, we can directly partition users into homogenous groups by considering the dominant topic of their observed preference profile:

$$cluster(u) = \underset{k}{\operatorname{argmax}} \ P(k|u).$$

For each user community, identified by a latent factor k, we can identify prototypical users by scoring them according to the measure:

$$score(u, k) = P(k|u)P(u|k).$$

Table 4.3: Top-10 characteristic movies for topic in Movielens1M—LDA(10)

Topic 1	Topic 2	Topic 3	Topic 4	Topic 5
Fly	American Beauty	Crying Game	Casablanca	Rock
Alien	Gladiator	Like Water for Chocolate	Chinatown	Total Recall
Rocky Horror Picture Show	Sixth Sense	Player	North by Northwest	Die Hard 2
Blade Runner	Austin Powers	Secrets and Lies	Maltese Falcon	Mission: Impossible
Alien 2	Mission: Impossible 2	Boys Don't Cry	Annie Hall	Independence Day
Aliens	Erin Brockovich	Ice Storm	African Queen	Hunt for Red October
Thing	American Pie	Dead Man Walking	Rear Window	True Lies
Twelve Monkeys	Being John Malkovich	Run Lola Run	The Graduate	Speed
Army of Darkness	The Patriot	Boogie Nights	Manchurian Candidate	Clear and Present Danger
Alien: Resurrection	X-Men	Big Night	Some Like It Hot	Fugitive

Topic 6	Topic 7	Topic 8	Topic 9	Topic 10
Rock	The Silence of the Lambs	Groundhog Day	Jerry Maguire	Beauty and the Beast
Total Recall	Fargo	Four Weddings and a Funeral	As Good As It Gets	Lion King
Die Hard 2	Star Wars V	Clueless	Titanic	Aladdin
Mission: Impossible	Star Wars IV	The Full Monty	Ghost	Snow White and the Seven Dwarfs
Independence Day	Shawshank Redemption	Shakespeare in Love	Forrest Gump	Little Mermaid
Hunt for Red October	Saving Private Ryan	Toy Story	American President	Fantasia
True Lies	Pulp Fiction	Forrest Gump	Speed	Lady and the Tramp
Speed	Raiders of the Lost Ark	There's Something About Mary	A Few Good Men	Cinderella
Clear and Present Danger	Schindler's List	When Harry Met Sally	Sleepless in Seattle	Peter Pan
Fugitive	American Beauty	Wayne's World	League of Their Own	Willy Wonka and the Chocolate Factory

Alternatively, the latent topics can devise an alternate feature space upon which to map each user. That is to say, we can exploit the predictive distribution $\boldsymbol{\theta}_u = \{\vartheta_{u,k}\}_{k=1,...,K}$ (where $\vartheta_{u,k} = P(k|u)$) as an alternative representation of u in the latent space of the K factors. Then, users can be clustered in communities by measuring their similarity within this latent space:

$$dist(u, v) = KL(\boldsymbol{\theta}_u \| \boldsymbol{\theta}_v) + KL(\boldsymbol{\theta}_v \| \boldsymbol{\theta}_u),$$

where

$$KL(\boldsymbol{\theta}_u \| \boldsymbol{\theta}_v) = -\sum_k \vartheta_{u,k} \log \frac{\vartheta_{u,k}}{\vartheta_{v,k}}.$$

Clusters can then be formed using standard similarity-based approaches.

Online Task Discovery.   Probabilistic topic models can be exploited in e-commerce scenarios, or, more generally, in situations where they exhibit a behavior partitioned into sessions. A typical example is when the users browses a website looking for a book on a specific subject. In such a case, the prediction of the underlying subject (represented by a latent factor here) is useful for a more appropriate presentation of the available catalog, as well as to facilitate the search.

Specifically, we assume that $\mathcal{I}_{\mathbf{R}}(u)$ can be partitioned into $\mathcal{S}^1(u), \ldots, \mathcal{S}^S(u)$, where $\mathcal{S}^j(u) \subseteq \mathcal{I}_{\mathbf{R}}(u)$ represents the j-th session in temporal order. Also, let $w(u, i, j)$ denote the number of times that the user u chose item i within $\mathcal{S}^j(u)$, and $\mathcal{S}^{<j}(u) = \bigcup_{j' < j} \mathcal{S}^{j'}(u)$ be the set of all the previous web browsing sessions associated with same user. For a given user session j, we can compute its distribution over topics by looking at the likelihood of the items in the current session [194]:

$$P(k|\mathcal{S}^j(u), u) = \frac{\prod_i w(u, i, j) P(i|k) P(k|\mathcal{S}^{<j}(u), u)}{\sum_{k'} \prod_i w(u, i, j) P(i|k') P(k'|\mathcal{S}^{<j}(u), u)},$$

where

$$P(k|\mathcal{S}^{<j}(u), u) = \frac{\prod_i \sum_{l' < j} j w(u, i, j') P(i|k) P(k)}{\sum_{k'} \prod_i \sum_{l' < j} j w(u, i, j') P(i|k') P(k')}.$$

4.2.2   PATTERN DISCOVERY USING CO-CLUSTERS

The probabilistic formulation mixture model provides a powerful framework for discovering hidden relationships between users and items. As shown above, such relationships can have several uses in users segmentation, product catalog analysis, etc. Those approaches can be further refined by considering a co-clustering structure, as proposed by approaches such as the FMM/BMM or the Bayesian hierarchical models. In general, given two different user clusters that group users who have shown a similar preference behavior, a co-cluster allows the identification of commonly rated items and categories for which the preference values are different. For example, two user communities might agree on action movies while completely disagreeing on romance movies. The identification of the topics of interest and their sequential patterns for each user community lead to an improvement of the quality of the recommendation list and provide the user with a more personalized view of the system.

Co-Cluster Analysis. Given a co-cluster $\langle k, l \rangle$, we can analyze the corresponding distribution of rating values to infer the preference/interest of the users belonging to the community k on an item of the category l. Figure 4.9 graphically shows a block mixture model with ten users clusters and nine item clusters built on the MovieLens dataset. A hard clustering assignment has been performed both on users and items: each user u has been assigned to the factor c such that $c = \text{argmax}_{k=1,\cdots,K} \; P(k|u)$. Symmetrically, each item i has been assigned to the cluster $d = \text{argmax}_{l=1,\cdots,L} \; P(l|i)$.

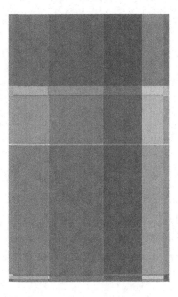

Figure 4.9: Blocks in the Movielens1M rating matrix.

The background color of each block $\langle k, l \rangle$ describes both the density of ratings and the average preference values given by the users (rows) belonging to the k-th group on items (columns) of the l-th category: the background intensity increases with the average rating values of the co-clusters, which are given in Table 4.4. Each point within the co-clusters represents a rating, and again, a higher rating value corresponds to a more intense color.

The analysis shows interesting tendencies: for example, users belonging to the user community c_1 tend to assign higher rating values than the average, and items belonging to item category d_6 are the most appreciated. A zoom of portions of the block image is given in Figure 4.10. Here, two blocks are characterized by opposite preference behaviors: the first block contains few (low) ratings, whereas the second block exhibits a higher density of high value ratings.

Table 4.4: Gaussian means for each block

	d_1	d_2	d_3	d_4	d_5	d_6	d_7	d_8	d_9
c_1	3.4	3.59	3.59	4	2.91	4.43	3.59	2.93	3.65
c_2	2.23	2.2	2.92	2.79	2	3.45	2.07	1.80	2.51
c_3	2.11	3.24	3	3.66	2	4.17	1	1.03	5
c_4	2.45	2.69	2.54	3.2	2.43	3.74	2.51	2	2.56
c_5	1	1.79	1	2.32	1	2.98	1.66	1	1.75
c_6	2.93	3.07	3	3.57	2.20	4.09	2.9	2.3	3.16
c_7	1	3.56	3.9	3.7	3.64	3.39	4	3.49	2
c_8	2.25	2.26	1.62	3.27	1	4.17	4.54	1	2.45
c_9	4.08	3.24	4.40	3.54	5	4	3.71	4.5	5
c_{10}	1.91	2.82	1	2.7	4.3	2.2	1	4	2

(a) Co-cluster (c_5, d_8): Mean rating 1   (b) Co-cluster (c_1, d_6): Mean rating 4.43

Figure 4.10: Co-cluster analysis.

The tasks of labeling and task discovery easily can be adapted to blocks representing simultaneous co-clusters. For example, we can exploit genres associated with movies to characterize item clusters, and, consequently, to associate a degree of interest of each community for the genres associated with each item cluster.

Modeling Topic Transitions. When preferences exhibit a temporal order, we can also detect sequential connections among topics by analyzing, e.g., which item categories are likely to next capture the interests of a user. Sequential patterns can be modeled by exploiting *Markov models*. The latter are probabilistic models for discrete processes characterized by the Markov properties. We adopt a Markov chain property here, i.e., a basic assumption that states that any future state only depends from the present state. This property limits the "memory" of the chain, which can be represented as a digraph where nodes represent the actual states and edges represent the possible transitions among them.

Assuming that the last observed item category for the considered user is d_i, the user could pick an item belonging to another topic d_j with probability $p(d_j|d_i)$. We shall focus on a more detailed treatment of Markov models in Chapter 6. For the moment, we provide a naive approach for the estimation of the *transition probabilities*, starting from a $|L + 1|$ x $|L + 1|$ *transition count matrix* $\mathcal{T}_c$, where $\mathcal{T}_c(i, j)$ stores the number of times that category j follows i in the rating profile of the users.[5]

The estimation we provide is rather simple, corresponding to a simple frequency count:

$$p(d_j|d_i) = \frac{\mathcal{T}_c(i, j)}{\sum_{j=1}^{L+1} \mathcal{T}_c(i, j')}.$$

Figure 4.11 represents the overall transition probability matrix, which highlights some strong connections among given categories. For example, with item categories having drama as the dominant genre, d_4, d_6, and d_9 are highly correlated, as well as d_2, d_7, and d_8, which correspond to comedy movies.

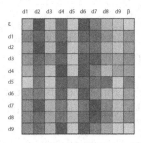

Figure 4.11: Transition probabilities matrix.

It is interesting to compare how the transition probabilities change within different user communities. Figure 4.12 shows the transitions for three different communities. Notice that, besides common transition patterns, each community has some distinctive transitions that characterize their population. For all the considered user communities, the most likely initial item category is d_6; while the first and the last community reproduced in the example show a strong attitude corresponding to the transition $d_8 \rightarrow d_2$, this is, instead, a weak pattern within c_7. The same consideration can be done for the transition $d_9 \rightarrow d_7$, which is strong for c_7 and c_{10}, while users belonging to c_3 are more prone to the transition toward d_6.

The analysis of the transition probabilities can be exploited for generating new recommendations enforcing topic diversity [211] in the top-K lists of items by taking into account not only the current topic of interest, but even the ones that are more likely to be connected to it.

4.2.3 DIVERSIFICATION WITH TOPIC MODELS

Given that the spectrum of the user interest can be wide, recommendations should ideally cover the latter as much as possible, by balancing accuracy and coverage in terms of abstract topics. This design principle becomes more important if we consider that recommendations and purchases

---

[5]We assume two further states ϵ, representing the initial choice, and β, representing the last choice.

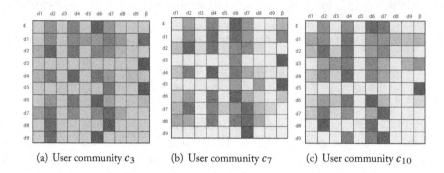

(a) User community c_3 (b) User community c_7 (c) User community c_{10}

Figure 4.12: Transition probabilities matrix.

can be performed by households. It is clear that each member of the family can exhibit different interests, and their actual identity is hidden by the system, which just records a unique user identifier.

Traditional approaches to recommendation often fail in providing diverse recommendations, as they tend to gather similar items in the same recommendation list. Unfortunately, this directly affects the novelty of recommendations: as items to be recommended can be selected by the ranking according to the highest mode of the user-specific distribution on interests/catalog categories, the final recommendation list will presumably contain very similar items. The lack of diversification enhances the *online filter bubble* [147], overplaying personalization, preventing concept drifts and the discovery of novel items.

Diversity rapidly became one of the most important dimensions of analysis and evaluation for recommendation techniques, attracting the attention of both academic and industry communities. The main challenge relies on the intrinsic evaluation of this measure. While prediction and recommendation accuracy metrics assess the accuracy at each individual entry, to evaluate diversity, we need to consider the whole recommendation list [126]. Assuming that a similarity function between items is available, one way of enforcing diversity is to re-rank items by promoting *intra-list dissimilarity* [211]. The goal of ensuring both diversity and accuracy for the recommendation lists provided by the system can be achieved by modeling this setting as a binary optimization problem [207].

Probabilistic techniques can be exploited to provide diverse recommendations. More specifically, we can promote diversity in two principled ways, which is discussed next.

Profile Partitioning.   The main idea, introduced in [208], is to partition the user profiles into clusters of similar items, and then generate the recommendation lists by selecting items that better match with each cluster. This approach can be nicely implemented by relying on probabilistic topic models for preference data modeling. While in Section 4.1.2 the item ranking function in case of topic models was obtained by marginalizing over all the topics, to enforce diversity, we can select a subset of topics $\mathcal{Z}_u \subseteq 2^K$ that best match the interests of the user u and then instantiate the ranking as:

$$p_i^u = \operatorname*{argmax}_{k \in \mathcal{Z}_u} P(i|k).$$

Explicit Diversification.   The recommendation list can be computed greedily, by iteratively selecting the item that provides the best compromise between the probability of being selected by the user and diversification with respect to the current list. Following [194], given an item i and a partial recommendation list $\mathcal{L}$, the *xQuaAD* objective function [172] for explicit diversification can be instantiated as follows:

$$\lambda\, P(i|u) + (1-\lambda) P(i,\overline{\mathcal{L}}|u),$$

where the first component promotes items that are likely to be selected by the user, while the second one is the likelihood of observing the item i but not the other items in the current list. The parameter $0 \le \lambda \le 1$ implements a tradeoff between item selection and diversity. Assuming that, given a topic k, the observation of i is independent of the items already selected, we can instantiate the latter component as:

$$P(i,\overline{\mathcal{L}}|u) = \sum_{k=1}^{K} P(k|u) P(i|k) \prod_{j \in \mathcal{L}} (1 - P(j|k)).$$

CHAPTER 5

Contextual Information

5.1 INTEGRATING CONTENT FEATURES

The previous chapters focused on collaborative filtering techniques, which generate recommendations that are based on the estimate of the correlation of historical rating data across the population of users/items. Content-based filtering is an alternative paradigm that has been used for recommending items, such as books, web pages, news, etc., that can be defined by informative content descriptors. It is natural to wonder whether the combination of collaborative and content filtering provides further advantages in terms of recommendation accuracy or additional application scenarios, and whether probabilistic modeling can easily integrate both of these sources of data. In a sense, collaborative and content-based filtering are complementary views that should be profitably exploited in a common learning architecture.

In a general setting, we can assume a function $\Psi : \mathcal{U} \times \mathcal{I} \times \mathcal{V} \mapsto \mathbb{R}^d$ that associates each preference observation $\langle u, i, r \rangle$ with a d-dimensional feature vector representing specific information associated with the preference observation. Some features may only depend on the underlying item (e.g., genre or category), or only on the user (such as demographic info like sex, city of residence, etc.), but, they may also combine aspects of both. The feature vector $\Psi(u, i, r)$ can hence be exploited to provide better recommendations.

There are several ways to cast the problem of predicting preference values when contextual information is available as a standard classification problem, or even as an ordinal regression problem. For example, the feature vector can be seen as a more general representation of the logistic features f_k introduced with the log-lineal model discussed in Chapter 2. Thus, the problem can be expressed in terms of logistic regression with feature weights $\mathbf{b} = \{\beta_1, \ldots, \beta_d\}$, and solved accordingly. Alternatively, one could fit a linear regression model, by assuming that the rating r is sampled from a Gaussian distribution centered at $\mathbf{b}^T \Psi(u, i, r)$ with variance fixed variance σ^2.

These basic models can be further enhanced to incorporate latent factor information. For example, in the case of explicit preference ratings, we can extend the basic PMF model shown in Chapter 3: recall that this model defines the probability of observing a given rating for a pair $\langle u, i \rangle$ as a Gaussian centered on $\mathbf{P}_u^T \mathbf{Q}_i$, where $\mathbf{P}$ and $\mathbf{Q}$ are the latent feature matrices. The extension to accommodate content features can be accomplished by still considering the rating value as the result of sampling from a Gaussian distribution, where, however, the center depends on both the latent factors $\mathbf{P}$ and $\mathbf{Q}$ and the feature weights $\mathbf{b}$:

$$P(r|u, i, \mathbf{b}, \mathbf{P}, \mathbf{Q}, \sigma) = \mathcal{N}(r; \mathbf{P}_u^T \mathbf{Q}_i + \mathbf{b}^T \Psi(u, i, r), \sigma^2).$$

The discrete case can be approached in a similar way, but it requires a more refined latent feature modeling. Here, $\mathbf{P}$ and $\mathbf{Q}$ represents three-dimensional tensors and the matrix $\mathbf{P}_r$ (resp. $\mathbf{Q}_r$) represents the set of latent features associated with rating r. Then, the probability of a specific rating r can be modeled as:

$$P(r|u, i, \mathbf{b}, \mathbf{P}, \mathbf{Q}, \sigma) = \frac{1}{Z_{u,i}(\mathbf{b}, \mathbf{P}, \mathbf{Q})} \exp \left\{ \mathbf{P}_{r,u}^T \cdot \mathbf{Q}_{r,i} + \mathbf{b}^T \Psi(u, i, r) \right\},$$

where $Z_{u,i}(\mathbf{b}, \mathbf{P}, \mathbf{Q})$ is a normalization factor defined as

$$Z_{u,i}(\mathbf{b}, \mathbf{P}, \mathbf{Q}) = \sum_{r \in \mathcal{V}} \exp \left\{ \mathbf{P}_{r,u}^T \cdot \mathbf{Q}_{r,i} + \mathbf{b}^T \Psi(u, i, r) \right\}.$$

Menon and Elkan [131] study the predictive abilities of such models for a wide range of applications, including recommendation on the Movielens dataset.

The idea behind the models presented so far is the starting point of more advanced approaches that include side information to achieve better results in prediction accuracy and provide tools for cold-start recommendation [3, 33, 187].

5.1.1   THE COLD-START PROBLEM

The exceptional sparseness of the rating matrix $\mathbf{R}$ poses serious challenges to the application of collaborative filtering techniques. In a real-world scenario, observations are available only for a small fraction of possible user/item pairs, and the distribution of observations is typically heavy-tailed (see Figure 4.1). Notably, a small fraction of users/items typically accounts for a large fraction of data, while the remaining are sparsely distributed among others.

Moreover, dynamic systems involve new users and/or items, and the prediction of preferences is required for these subjects as well. The problem of providing predictions for users/items with little, or no, historical preference information, is commonly referred to as a *cold-start problem* (see Chapter 1). By contrast, *warm-start* refers to cases in which recommendations must be provided to users/items for which a sufficient number of past preferences in the rating matrix $\mathbf{R}$ is available.

Although effective for warm-start, collaborative filtering approaches fail to address the cold-start problem. Hence, methods based on the idea shown above, which simultaneously incorporate feedback data and user/item features, can provide accurate predictions and smoothly handle both cold-start and warm-start scenarios.

Following [2], it is convenient to specify $\Psi(u, i, r)$ in three components, namely $\Psi(u, i, r) = (\mathbf{v}_u, \mathbf{w}_i, \mathbf{s}_{u,i})$. Here, $\mathbf{v}_u \in \mathbb{R}^p$ represents the user-specific features, $\mathbf{c}_i \in \mathbb{R}^q$ represents item-specific features, and $\mathbf{s}_{u,i} \in \mathbb{R}^s$ represents features relative to the dyad. Then, we can refine the predictive model by assuming the existence of latent factors $\{\alpha, \beta, \mathbf{P}, \mathbf{Q}\}$ and regression weights $\mathbf{b}$, which govern the conditional probability of observing a rating value r associated to the pair $\langle u, i \rangle$:

(Continuous rating) $r \sim \mathcal{N}(m_{u,i}, \sigma^2)$, where

$$m_{u,i} = \mathbf{s}_{u,i}^T \mathbf{b} + \alpha_u + \beta_i + \mathbf{P}_u^T \mathbf{Q}_i;$$

(Discrete rating) $r \sim Disc(\{m_{u,i,r}\}_{r \in \mathcal{V}})$, where

$$m_{u,i,r} \propto \exp\{\mathbf{s}_{u,i}^T \mathbf{b} + \alpha_u + \beta_i + \mathbf{P}_{r,u}^T \mathbf{Q}_{r,i}\}.$$

All latent factors can be expressed in terms of further probabilistic regression setting, as specified by the following regression equations:

$$\begin{aligned}
\alpha_u &= \mathbf{g}_0^T \mathbf{v}_u + \epsilon_u^\alpha, & \epsilon_u^\alpha &\sim \mathcal{N}(0, a_\alpha); \\
\beta_i &= \mathbf{d}_0^T \mathbf{c}_i + \epsilon_i^\beta, & \epsilon_i^\beta &\sim \mathcal{N}(0, a_\beta); \\
\mathbf{P}_u &= \mathbf{G}\mathbf{v}_u + \epsilon_u^U, & \epsilon_u^U &\sim \mathcal{N}(0, \Sigma_U); \\
\mathbf{Q}_i &= \mathbf{D}\mathbf{c}_i + \epsilon_i^V, & \epsilon_i^V &\sim \mathcal{N}(0, \Sigma_V).
\end{aligned}$$

The *Linear Regression Factor Model* (LRFM, [2]) implements this setting by assuming a two-stage generative process. In the first stage latent factors are sampled, while in the second stage latent factors are exploited to generate rating values:

- For each $\langle u, i \rangle$:

  1. Sample the factors

$$\begin{aligned}
\alpha_u &\sim \mathcal{N}(\mathbf{g}_0^T \mathbf{v}_u, a_\alpha), & \beta_i &\sim \mathcal{N}(\mathbf{d}_0^T \mathbf{c}_i, a_\beta), \\
\mathbf{P}_u &\sim \mathcal{N}(\mathbf{G}\mathbf{v}_u, \Sigma_U), & \mathbf{Q}_i &\sim \mathcal{N}(\mathbf{D}\mathbf{c}_i, \Sigma_V),
\end{aligned}$$

  and compute $m_{u,i} = \mathbf{s}_{u,i}^T \mathbf{b} + \alpha_u + \beta_i + \mathbf{P}_u^T \mathbf{Q}_i$;

  2. Sample the rating r:

$$r \sim \mathcal{N}(m_{u,i}, \sigma^2).$$

The graphical model for LRFM, depicted in Figure 5.1, shows strong similarities with the PMF model described in Chapter 2. And in fact, by simply enforcing the hyperparameters $\mathbf{G}, \mathbf{D}, \mathbf{g}_0$, and $\mathbf{d}_0$ to zero we simplify the model to the standard PMF. Furthermore, if we assume, instead, zero variance on factors, by assuming a_α, a_β and the covariance matrices Σ_U and Σ_V to be zero, the resulting model simplifies to a standard regression model based on content features.

LRFM represents a robust combination of both content and collaborative filtering based on factor modeling and linear regression. As a result, the model is capable of handling both cold-start and warm-start in a uniform way.

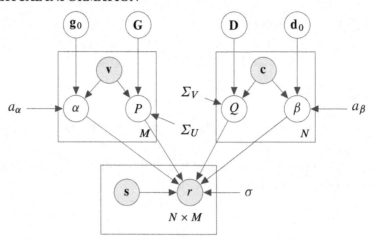

Figure 5.1: Graphical model for LRFM.

The complete data likelihood for a given set $\mathcal{X}$ of observations and latent factors $\Theta = \{\alpha, \beta, \mathbf{P}, \mathbf{Q}\}$ can be expressed as

$$P(\mathcal{X}, \Theta | \mathbf{D}, \mathbf{G}, \mathbf{g}_0, \mathbf{d}_0) = \prod_{\langle u,i,r \rangle} P(r | \Theta, \mathbf{v}_u, \mathbf{c}_i, \mathbf{s}_{u,i}) P(\alpha_u | \mathbf{v}_u, \mathbf{g}_0) P(\beta_i | \mathbf{c}_i, \mathbf{d}_0)$$
$$P(\mathbf{P}_u | \mathbf{v}_u, \mathbf{G}) P(\mathbf{Q}_i | \mathbf{c}_i, \mathbf{D}). \tag{5.1}$$

This joint likelihood can be maximized by resorting to the Stochastic EM procedure (described in Section A.3), where the E-step corresponds to sampling the Θ parameters by means of Gibbs sampling. Posterior probabilities for the latent factors can be expressed in terms of Gaussian distribution, thus enabling a simple Gibbs step. In addition, the M step consists of finding the optimal hyperparameters $\mathbf{D}, \mathbf{G}, \mathbf{g}_0, \mathbf{d}_0$ given $\mathcal{X}$ and the Θ sampled in the E-step. Again, since the underlying distributions are Gaussians, the resulting estimation procedure is straightforward (see [2] for details).

5.1.2 MODELING TEXT AND PREFERENCES

A natural evolution of the modeling proposed so far is to consider the case where each item can be characterized by meta data available as free text. Textual descriptions are a common in several scenarios, e.g. content recommendation, and hence they can provide useful information that adds up to other features naturally represented as vectors in a Euclidean space.

Furthermore, we can consider highly dynamic scenarios, such as recommendation of news articles, where the notion of item is not even static. In these settings, the set $\mathcal{I}$ can be considered as stream and continuous bursts of new items that need to be appropriately handled; hence, it is convenient to represent each item through a set of features, and measure how these features affect the evaluation by users. Textual features in this respect may represent the most significant informative content that can be associated with items.

When dealing with text, it is convenient to abstract its representation to a set of latent features. Given the bag-of-words representation $\mathbf{w}_i$ corresponding to the item i, we can assume that the vector $\mathbf{z}_i$ represents the set of latent topics, where $z_{i,j}$ is the latent topic associated with the word $w_{i,j}$. By the term "words," we denote any elementary information, such as textual tokens, entities, etc.

In a matrix factorization setting, the rating associated to a dyad $\langle u, i \rangle$ can be modeled by resorting to latent factors $\boldsymbol{\alpha}, \boldsymbol{\beta}, \mathbf{P}, \mathbf{Q}$. In particular, $\mathbf{Q}_i$ can be interpreted as a representation of $\mathbf{w}_i$ within the latent space. However, since $\mathbf{w}_i$ is associated with $\mathbf{z}_i$, it is natural to enforce a dependency between $\mathbf{Q}_i$ and $\mathbf{z}_i$. Agarwal and Chen in [3] reformulate $\mathbf{Q}_i$ as the vector $\tilde{\mathbf{z}}_i$. The latter, in turn, depends on the topic assignments within $\mathbf{z}_i$ as follows:

$$\tilde{z}_{i,k} = \frac{1}{|\mathbf{w_i}|} \sum_j \mathbb{1}[\![z_{i,j} = k]\!].$$

This reformulation leads to the definition of the *fLDA* model, shown graphically in Figure 5.2, and it is a variation of LRFM. Notice that the left-hand side of the graphical model is identical to LRFM. What differs is the right-hand side, where item observations are modeled. This component corresponds to an LDA model, where the sampled topics are exploited upon the generation of the rating:

- For each $\langle u, i \rangle$:

    1. Sample the factors

    $$\alpha_u \sim \mathcal{N}(\mathbf{g}_0^T \mathbf{v}_u, a_\alpha), \qquad \beta_i \sim \mathcal{N}(\mathbf{d}_0^T \mathbf{c}_i, a_\beta), \qquad \mathbf{P}_u \sim \mathcal{N}(\mathbf{G}\mathbf{v}_u, \Sigma_U);$$

    2. For $j = \{1, \ldots, |\mathbf{w}_i|\}$:
        (a) sample $z_{i,j} \sim Dir(\gamma)$;
        (b) sample $w_{i,j} \sim Disc(\Phi_{z_{i,j}})$;
    3. Compute $m_{u,i} = \mathbf{s}_{u,i}^T \mathbf{b} + \alpha_u + \beta_i + \mathbf{P}_u^T \mathbf{z}_i$;
    4. Sample $r \sim \mathcal{N}(m_{u,i}, \sigma^2)$.

Parameter estimation can be accomplished as a variation of the Stochastic EM procedure discussed above. In practice, the E step consists of a Markov chain where the components α, β and $\mathbf{P}, \mathbf{Z}$ are iteratively sampled.

Rather than embedding the LDA annotation within the matrix factorization process, it is also possible to consider a direct extension of the URP model shown in Chapter 3, which accounts for a textual description of the item. Following the standard topic modeling approach,

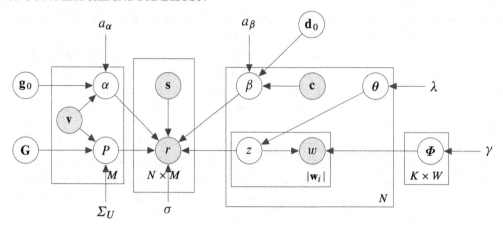

Figure 5.2: Graphical model for fLDA.

each preference observation is associated with a topic $k \in \{1, \cdots, K\}$. However, the rating distribution $\phi_{k,i}$ depends upon the topic assignments relative to the underlying textual description $\mathbf{w}_i$. The generative process can be described as follows:

- For each $u \in \mathcal{U}$ sample $\boldsymbol{\theta}_u \sim Dir(\boldsymbol{\alpha})$.

- For each $i \in \mathcal{I}$ sample $\boldsymbol{\vartheta}_i \sim Dir(\boldsymbol{\lambda})$.

- For each $k \in \{1, \ldots, K\}$ sample $\boldsymbol{\Phi}_k \sim Dir(\boldsymbol{\beta})$.

- For each $\langle u, i \rangle$:

  1. Sample $z_u \sim Disc(\boldsymbol{\theta}_u)$;
  2. For $j = \{1, \ldots, |\mathbf{w}_i|\}$:
     (a) sample $z_{i,j} \sim Disc(\boldsymbol{\vartheta}_i)$;
     (b) sample $w_{i,j} \sim Disc(\boldsymbol{\Phi}_{z_{i,j}})$;
  3. Compute $\mathbf{m}_{u,i} = \frac{1}{n_i} \sum_j \mu_{z_u, z_{i,j}}$;
  4. Sample $r \sim \mathcal{N}(\mathbf{m}_{u,i}, \sigma^2)$.

5.2 SEQUENTIAL MODELING

All probabilistic approaches studied so far propose a data generation process that is based on the bag-of-words assumption, i.e., such that the temporal order of the items accessed by a user can be neglected and co-occurrence patterns are used to define the topic space. However, an alternative view is that traces can be "naturally" interpreted as sequences where the temporal order is relevant. Ignoring the intrinsic sequentiality of the data may result in poor modeling. So far, all observations were assumed to be independent and identically distributed. On the other hand, sequential data may express causality and dependency, and different factors can be used to characterize different dependency likelihoods. The focus here is the context in which the user acts and expresses preferences, i.e., the environment, characterized by side information, where observations are recorded.

The analysis of sequential patterns has important applications in modern recommender systems, especially when we are interested in achieving a balance between personalization and contextualization. For example, in Internet-based streaming services for music or video (such as Last.fm[1] and Videolectures.net[2]), the context of the user interaction with the system easily can be interpreted by analyzing the content previously requested. The assumption here is that the current item (and/or its genre) influences the next choice of the user. In particular, if a specific user is in the "mood" for classical music (as observed in the current choice), it is unlikely that the immediate subsequent choice will be a song of a completely different genre. Being able to capture such properties and exploit them in recommendation strategy can greatly improve the accuracy of the recommendation.

5.2.1   MARKOV MODELS

Sequential correlation between observations can be modeled by relaxing the i.i.d. assumption, i.e., by enforcing a dependency between the current observation and the previous ones. In probabilistic terms, we can assume a temporal ordering $\mathbf{x}_1 \ldots, \mathbf{x}_N$ of observations in $\mathcal{X}$, and then express the probability by using the chain rule

$$P(\mathcal{X}|\Theta) = P(\mathbf{x}_1) \prod_{i=2}^{N} P(\mathbf{x}_i|\mathbf{x}_1, \ldots, \mathbf{x}_{i-1}, \Theta).$$

Within this formula, a sequence can be modeled as a stationary *first-order Markov chain*.

- A Markovian process naturally models the sequential nature of the data, where dependencies among past and future tokens reflect changes over time that are still governed by similar features;

- the chain is stationary, as a fixed number of tokens is likely to appear frequently in sequences, and their joint distribution is invariant with respect to shifts in time;

- the order of the chain is 1 because the possibility that two subsequent tokens share some features is more likely than that of two tokens distant in time.[3]

---

[1]http://last.fm
[2]http://videolectures.net
[3]It is also worth noticing that higher-order dependencies introduce an unpractical computational overhead, as the number of parameters grows exponentially with the order of the chain [31, Chapter 13].

The simplest way to relax the i.i.d. assumption is to assume that the current observation is independent from all the previous observations except the most recent, i.e.,

$$P(\mathcal{X}|\Theta) = P(\mathbf{x}_1) \prod_{i=2}^{N} P(\mathbf{x}_i|\mathbf{x}_{i-1}, \Theta).$$

A first order Markov chain represents a trade-off between complexity and expressiveness. For example, we can consider a Markovian extension of the multinomial mixture model shown in Figure 2.2, where the dependency is made explicit: in this model (depicted graphically in Figure 5.3), we can assume a latent factor associated with a user. Such a latent factor defines a parameter set $\boldsymbol{\Phi}$ representing the probability of observing a rating given a previous rating. That is, $\phi_{k,r_1.r_2}$ represents the (multinomial) probability of observing r_2, given that r_1 was observed before and the latent factor is k.

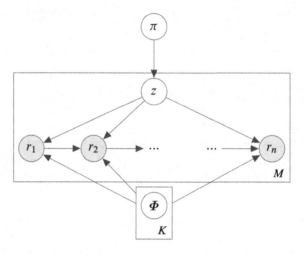

Figure 5.3: Graphical model for the Markov Mixture Model.

Several variations of this simple model can be proposed. For example, if we restrict our attention to implicit preferences only, we can assume that $\boldsymbol{\Phi}$ models item selection, so that $\phi_{k,i.j}$ represents the probability that a user associated with topic k will select item j, given that her previous choice was i. This model was proposed in [37], to analyze the navigation behavior of users in a website.

In the following, we will still consider implicit preferences, and we focus on extensions of the basic LDA framework that exploit Markovian dependencies. These extensions have been studied in the context of text modeling and natural language processing. In [195, 199], for example, authors propose extensions of the LDA model that assume a first-order Markov chain for the word generation process. In the resulting *Token-Bigram Model* and *Topical n-grams*, the current word depends on the current topic and the previous word observed in the sequence. We shall see how such paradigms can be exploited for recommendation.

Bayesian Modeling

The starting point is to reconsider the general formulation of the LDA model. Figure 5.4(a) provides an alternative graphical view of the latter, where all the selected items are unfolded.

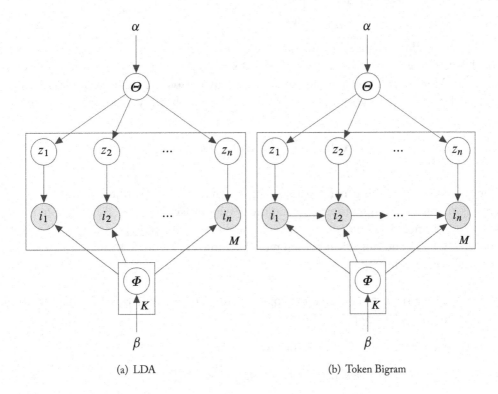

(a) LDA   (b) Token Bigram

Figure 5.4: LDA and token bigram.

If we consider the observations $\mathcal{X}$ as sequences, where each sequence $\mathbf{x}_u \triangleq w_{u,1} \cdot w_{u,2} \cdot \ldots \cdot w_{u,n_u}$ is relative to a user u and $w_{u,j}$ is a token representing the $j-th$ item chosen by u in temporal order, $\mathbf{Z}$ encodes the latent topics relative to each selection, while $\boldsymbol{\Phi}$ and $\boldsymbol{\Theta}$ are the distribution functions governing the likelihood of $\mathcal{X}$ and $\mathbf{Z}$ (with respective priors β and α). The complete likelihood can be expressed as:

$$P(\mathcal{X}, \mathbf{Z}, \boldsymbol{\Theta}, \boldsymbol{\Phi} | \alpha, \beta) = P(\mathcal{X}|\mathbf{Z}, \boldsymbol{\phi}) P(\boldsymbol{\phi}|\beta) P(\mathbf{Z}|\theta) P(\theta|\alpha),$$

where

$$P(\mathcal{X}|\mathbf{Z},\boldsymbol{\phi}) = \prod_{u\in\mathcal{U}} P(\mathbf{x}_u|\mathbf{z}_u,\boldsymbol{\phi}) \qquad P(\mathbf{Z}|\boldsymbol{\theta}) = \prod_{u\in U} P(\mathbf{z}_u|\boldsymbol{\theta}_u), \tag{5.2}$$

and $P(\boldsymbol{\Theta}|\alpha)\,P(\boldsymbol{\Phi}|\beta)$ represent the usual Dirichlet priors. In the standard LDA setting where all tokens are independent and exchangeable, the two components can be specified as

$$P(\mathbf{x}_u|\mathbf{z}_u,\boldsymbol{\Phi}) = \prod_{m\equiv\langle u,i\rangle} \varphi_{z_m,i} \qquad P(\mathbf{z}_u|\boldsymbol{\theta}_u) = \prod_{m\equiv\langle u,i\rangle} \vartheta_{u,z_m}. \tag{5.3}$$

Relaxing the exchangeability assumptions in Equations 5.3 on either $\mathbf{x}_u$ or $\mathbf{z}_u$ allows us to model different sequential dependencies. A first straightforward extension is to consider $\mathbf{x}_u$ as a first-order Markov chain, where each token $w_{u,j}$ depends on the most recent token $w_{u,j-1}$ observed so far. This model was proposed in [195], and the probability of observing a trace becomes:

$$P(\mathbf{x}_u|\mathbf{z}_u,\phi) = \prod_{j=1}^{N_d} P(w_{u,j}|w_{u,j-1},z_{u,j},\phi). \tag{5.4}$$

In practice, a token $w_{u,j}$ is generated according to a multinomial distribution $\phi_{z_{u,j},w_{u,j-1}}$, which depends on both the current topic $z_{u,j}$ and the previous token $w_{u,j-1}$.[4] According to this model, the conjugate prior β accounts for all the possible distributions $\{\phi_{k,i}\}_{k=1,\ldots,K;i\in\mathcal{I}}$. The difference between the LDA and this token bigram model is shown in Figure 5.4, and the generative model can be described as follows.

- For each user $u \in \mathcal{U}$ sample the topic-mixture components $\theta_d \sim Dir(\vec{\alpha})$ and sequence length $n_d \sim Poisson(\xi)$.

- For each topic $k \in \{1,\ldots,K\}$ and token $i \in \mathcal{I}$;

  - Sample token selection components $\phi_{k,i} \sim Dir(\beta)$.

- For each user $u \in \mathcal{U}$ and $j \in \{1,\ldots,n_u\}$;

  - sample a topic $z_{u,j} \sim Disc(\theta_u)$;

  - sample a token $w_{u,j} \sim Disc(\phi_{z_{u,j},w_{u,j-1}})$.

An important aspect to consider is the choice for the Dirichlet prior β. In a general form, we can assume a family $\beta \triangleq \{\beta_{k,i}\}_{k\in\{1,\ldots,K\};i\in\mathcal{I}}$ of Dirichlet coefficients. As shown in [195], different modeling strategies (e.g., shared priors $\beta_{k,r.s} = \beta_s$) can affect the accuracy of the model.

---

[4]When $j=1$, the previous token is empty and the multinomial resolves to $\vec{\phi}_{z_{u,j}}$, representing the initial status of a Markov chain.

Plugging the above probability within the likelihood equation allows us to exploit the standard algebraic manipulation already studied for the LDA and its variants, so that the inference phase can be obtained by means of the well-known stochastic EM strategy, where the E step consists of a collapsed Gibbs sampling procedure for estimating $\mathbf{Z}$, and the M step estimates both the predictive distributions θ and ϕ and the hyperparameters α and β given $\mathbf{Z}$. Within Gibbs sampling, topics are iteratively sampled, according to the probability:

$$P(z_{u,j} = k | \mathbf{Z}_{-(u,j)}, \mathcal{X}) \propto \left(n_{u,(\cdot)}^{k} + \alpha_k - 1 \right) \cdot \frac{n_{(\cdot),r.s}^{k} + \beta_{k,r.s} - 1}{\sum_{s' \in \mathcal{I}} n_{(\cdot),r.s'}^{k} + \beta_{k,r.s'} - 1}, \qquad (5.5)$$

relative to the topic, to associate with the j-th token of u's trace, where $w_{u,j-1} = r$ and $w_{u,j} = s$.
Given $\mathbf{Z}$, the parameters θ and ϕ can be estimated according to the following equations:

$$\vartheta_{u,k} = \frac{n_{u,(\cdot)}^{k} + \alpha_k}{\sum_{k'=1}^{K} (n_{u,(\cdot)}^{k'} + \alpha_{k'})} \qquad \varphi_{k,r.s} = \frac{n_{(\cdot),r.s}^{k} + \beta_{k,r.s}}{\sum_{s' \in \mathcal{I}} (n_{(\cdot),r.s'}^{k} + \beta_{k,r.s'})}. \qquad (5.6)$$

These distributions can be directly exploited for recommendation purposes, by making explicit the dependency of the current selection from the previous. By denoting with $rank(i, u)$ the degree of interest that user u might express for i, we have:

$$rank(i, u) = \sum_{k=1}^{K} P(i | z_{u,n_u+1} = k, \mathbf{x}_u) P(z_{u,n_u+1} = k | \vec{\theta}_d) = \sum_{k=1}^{K} \varphi_{k,r.i} \cdot \vartheta_{u,k},$$

where $r = w_{u,n_u}$ is the last item selected by user u.

Hidden Markov Models

The bigram modeling can be extended by considering different kind of dependencies among hidden states of the model. In particular, we can assume that the dependencies can occur at a hidden state. *Hidden Markov models* (HMMs) [31, Chapter 13] are a general reference framework both for modeling sequence data and for natural language processing [122]. HMMs assume that sequential data is generated by a Markov chain of latent variables, with each observation conditioned on the state of the corresponding latent variable. The resulting likelihood can be interpreted as an extension of a mixture model, where the choice of mixture components for each observation is not selected independently but depends on the choice of components for the previous observation.

By looking again at the LDA model in Figure 5.4(a), we can enforce dependency between topics, rather than tokens. In Figure 5.5 tokens are considered independent of each other and related to a latent topic, while a topic depends on the previous ones. However, since topics represent the ultimate factors underlying a token appearance in the sequence, correlation between topics can better model an evolution of the underlying themes. Assuming a first-order Markovian dependency, the *topic bigram* model refines the Equation 5.3 by modeling the probability of

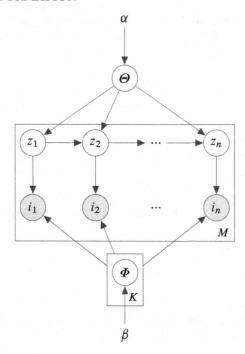

Figure 5.5: Topic bigram.

a sequence of latent topics as:

$$P(\mathbf{z}_u|\boldsymbol{\theta}_u) = \prod_{j=1}^{n_u} P(z_{u,j}|z_{u,j-1}, \boldsymbol{\theta}_u). \tag{5.7}$$

The difference here relies on the distribution generating $z_{u,j}$, which is a multinomial $\boldsymbol{\theta}_{u,z_{u,j-1}}$ parameterized by both u and the previously sampled topic $z_{u,j-1}$. Again, the generative process follows straightforwardly.[5]

- For each user $u \in \mathcal{U}$ and topic $h \in \{0, \ldots, K\}$ sample topic-mixture components $\boldsymbol{\theta}_{u,h} \sim Dir(\boldsymbol{\alpha})$ and sequence length $n_u \sim Poisson(\xi)$.

- For each topic $k = 1, \ldots, K$ sample token selection components $\varphi_k \sim Dir(\beta)$.

- For each $u \in \mathcal{U}$ and $j \in \{1, \ldots, n_u\}$ sequentially:

  - sample a topic $z_{u,j} \sim Disc(\boldsymbol{\theta}_{u,z_{u,j-1}})$;
  - sample a token $w_{u,j} \sim Disc(\phi_{z_{u,j}})$.

---

[5]Here, $h = 0$ is a special topic that precedes the first topic of each trace.

By algebraic manipulations we can obtain a closed formula for the joint distribution $P(\mathcal{X}, \mathbf{Z}|\alpha, \beta)$, which is the basis for a stochastic EM procedure based on a Gibbs sampler. The ranking function in the topic bigram model can be obtained by adapting the forward-backward algorithm (see [31, Section 13.2] for details). It is necessary to build a recursive chain of probabilities, representing a hypothetical random walk among the hidden topics:

$$rank(i, u) = \sum_{k=1}^{K} P(w_{u,n_u+1} = i, z_{u,n_u+1} = k|\mathbf{x}_u)$$

$$\propto \sum_{k=1}^{K} P(w_{u,1} \cdots w_{u,n_u+1} = i, z_{u,n_u+1} = k),$$

which requires solving $P(w_{u,1} \cdots w_{u,n_u+1}, z_{u,n_u+1})$. By algebraic manipulations:

$$P(w_{u,1} \cdots w_{u,N_d}, z_{u,n_u} = k) = P(w_{u,1} \cdots w_{u,N_u}|z_{u,N_u} = k)P(z_{u,n_u} = k)$$

$$= \varphi_{k,w_{u,n_u}} \sum_{h} P(w_{u,1} \cdots w_{u,N_u-1}, z_{u,N_u-1} = h)\vartheta_{u,h.k}.$$

The result is a recursive equation that can be simplified into the following γ function:

$$\gamma_k(\mathbf{x}_u; 1) = \varphi_{k,w_{u,1}}; \qquad \gamma_k(\mathbf{x}_u; j) = \varphi_{k,w_{u,j}} \sum_{h} \gamma_h(\mathbf{x}_u; j-1)\vartheta_{u,h.k}.$$

Therefore, the ranking function can be formulated as:

$$rank(i, u) \propto \sum_{k=1}^{K} \gamma_k(\mathbf{x}_u; n_u + 1).$$

There are several variations of the basic token and topic bigram models. [76] propose an *Hidden Topic Markov Model (HTMM)* for text documents. HTTM defines a Markov chain over latent topics of the document. The corresponding generative process, depicted in Figure 5.6(a), assumes that all words in the same sentence share the same topic, while successive sentences can either rely on the previous topic, or introduce a new one. Topics in a document form a Markov chain with a transition probability that depends on a binary topic transition variable ψ. When $\psi = 1$, a new topic is drawn for the n-th sentence, otherwise the same previous topic is used. *HTMM* models topic cohesion at the level of single tokens (e.g., words within the same sentence share the same latent topic), but does not model directly a smooth evolution between topics in different segments that frame a trace. *Sequential LDA* [54] is a variant of LDA which models a sequential dependency between sub-topics: the topic of the current segment is closely related to the topic of its antecedent and subsequent segments. This smooth evolution of the topic flow is modeled by using a Poisson-Dirichlet process.

The *LDA Collocation Model* [74] introduces a new set of random variables (for bigram status) x, which denotes whether a bigram can be formed with the previous token. More specifically, as represented in Figure 5.6(b), the generative process specifies for each word both a topic and a collocation status.

- if $x_i = 1$ then w_i is part of a collocation and thus is generated by sampling from a distribution conditioned on the previous word $P(w_i|w_{i-1}, x_i = 1)$;

- otherwise, w_i is sampled from a distribution associated to the current topic $P(w_i|z_i, x_i = 0)$.

The collocation status adds a more flexible modeling than the token bigram model, which always generates bigrams and, according to this formulation, the distribution on bigram does not depend on the topic.

Finally, the *token–bitopic model* [22] assumes that tokens are still related to past events. However, the assumption here is that a token depends on a chain of sequential topics, rather than a single topic.

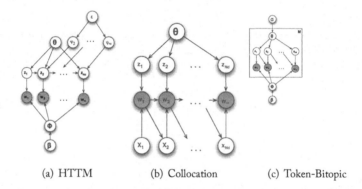

(a) HTTM   (b) Collocation   (c) Token-Bitopic

Figure 5.6: HTMM, collocation, and bitopic graphical model for the generation of a document.

5.2.2 PROBABILISTIC TENSOR FACTORIZATION

The sequential evolution of preferences and rating patterns can also be interpreted in terms of "seasonality" of preference values. Throughout this book, we have seen how latent factors are a powerful tool for capturing the correlations among users and items in a probabilistic settings. However, correlations can change within time intervals: a user might be interested in sci-fi books in the evening, whereas, during the day, the same user can be interested in technical books on computer science. In this case, trying to explain all the data with one fixed global model would be ineffective. On the other hand, concentrating on the recent choices still does not allow us to capture a comprehensive behavioral model for the same user.

This notion of sequentiality can be captured by extending the PMF models shown in Chapters 2 and 3. In addition to the factors that are used to characterize users and items, we can also introduce factors for time periods. Intuitively, these factors are associated with a given time period, and can represent a correlations among the given time periods and the preferences that users can express. This approach is discussed in [200] and it is called *probabilistic tensor factorization*.

Formally, we can assume that observations in $\mathcal{X}$ still represent the preference r of user u for item i, but we also assume that preferences can change over time intervals, so that a tuple $\langle u, i, r, t \rangle$ represents the preference at interval t. We also assume that intervals are finite and they represent seasonal time frames, i.e., $t \in \{1, \ldots S\}$. The tensor factorization model assumes the existence of the latent matrices $\mathbf{P} \in \mathbb{R}^{M \times K}$, $\mathbf{Q} \in \mathbb{R}^{N \times K}$, and $\mathbf{T} \in \mathbb{R}^{S \times K}$, so that a rating can be expressed as:

$$P(r|u, i, t, \mathbf{P}, \mathbf{Q}, \mathbf{T}, \sigma) = \mathcal{N}(r; \sum_k \mathbf{P}_{u,k} \mathbf{Q}_{i,k} \mathbf{T}_{t,k}, \sigma^2).$$

The interpretation of the above factorization is that a rating depends not only on how similar a user's preferences and an item's features are (as in PMF), but also on how much these preferences/features match with the current trend as reflected in the time feature vectors.

As in the PMF framework, the $\mathbf{P}$ and $\mathbf{Q}$ factor matrices can be regularized by imposing Gaussian priors, i.e.,

$$\mathbf{P}_u \sim \mathcal{N}(\boldsymbol{\mu_u}, \Sigma_U), \qquad \mathbf{Q}_i \sim \mathcal{N}(\boldsymbol{\mu_I}, \Sigma_I).$$

As for the $\mathbf{T}$ matrix, we can assume that the seasonality can be expressed through a smooth dependency between adjacent time intervals. That is to say, the values in $\mathbf{T}_t$ depend on the values of its predecessor $\mathbf{T}_{t-1}$:

$$\mathbf{T}_1 \sim \mathcal{N}(\boldsymbol{\mu_0}, \Sigma_0), \qquad \mathbf{T}_t \sim \mathcal{N}(\mathbf{T}_{t-1}, \Sigma_T).$$

Inference and parameter estimation can be solved in a full Bayesian framework as in [168]: hyperparameters and the factor matrices are iteratively sampled according to a Gibbs sampling procedure that exploits appropriate conjugate priors to express the posteriors.

CHAPTER 6

Social Recommender Systems

The evolution of the Web and information technologies has made it possible to broaden the traditional concept of the diffusion of information, introducing a virtual environment where one can exchange ideas, opinions, and information. The *Social Web* realizes such ideas, by allowing different kinds of interactions among people with similar tastes. Social contents/relationships and microblogging features, such as following/follower relationships and sharing of "memes"[1] or profile updates, are quickly reshaping the idea of the Web, which is progressively moving from the original concept of connecting documents/resources to the idea of connecting people. This is witnessed by the incredible growth of *Social Networks* (SNs), both in terms of active accounts and user-generated content:

- In the first quarter of 2013, FaceBook[2] got over 1 billion monthly active users;[3]

- Google+[4] registered over 500 million accounts;[5]

- Twitter[6] produced over 9,000 tweets per second.[7]

SNs provide a new kind of complex and heterogenous data repository that can consistently improve the quality of a recommendation system by allowing a more accurate and fine-grained profiling of trends and users' tastes. For instance, ratings or opinions shared by social peers can be exploited to mitigate the sparsity of the user-item preference matrix. In a sense, our first source of recommendations is the social environment: we naturally ask friends for recommendations about the next movie to watch, or restaurants to frequent. In a comparison between social filtering and information filtering algorithms, recommendations from friends have been reported [184] to be consistently better than those from online recommender systems.

On the other hand, a good recommender system can enhance the experience of the users of the SN. The extremely dynamic stream of information and resources available in online networks represents a serious case of information overload. Information, memes, and opinions quickly spread across the network and produce new information, in an endless cycle. With the huge

---

[1]A meme refers generally to short snippets of text, photos, audio, or videos.
[2]http://www.facebook.com/
[3]http://investor.fb.com/releasedetail.cfm?ReleaseID=761090
[4]http://plus.google.com/
[5]http://googleblog.blogspot.it/2012/12/google-communities-and-photos.html
[6]http://www.twitter.com/
[7]http://www.statisticbrain.com/twitter-statistics/

amount of available products/services/information in the SNs, recommendation techniques play an important role in the detection of potentially interesting and attractive content for each user.

Social rating networks, which combine advantages of SNs and RSs, are emerging as collaborative platforms on which users discover and share opinions about items (movies, books, music, etc.). Examples are:

- Last.fm,[8] halfway between a music recommendation service and a social network.

- Flixster,[9] a social network where users share film reviews and ratings.

- The recent integration of TripAdvisor[10] on popular third-party social networking sites, including FaceBook, allows users to search for opinions about restaurants, hotels, and places to visit, shared by their social connections.

The integration of social features and data into recommender systems offers great opportunities, but it also calls for novel recommendation models that deal with two main problems. First of all, almost all current recommender systems are designed for specific domains and applications without explicitly addressing the heterogeneity of the implicit and explicit preference information available on SNs. Users' feedback on items may come in different format: implicit clicks, thumbs up/down, textual reviews, or numerical ratings. Moreover, some contextual features, such as location, time, or level of social engagement, can have varying importance on different recommendation domains. Finally, traditional recommender systems assume that all the users are independent and identically distributed, thus ignoring the existence of complex relationships between social interaction and similarity [46].

Online SNs exhibit an interesting property: users tend to form social ties with people having similar sociodemographic attributes, or tastes. In other words: "similarity breeds connection" [127]. This property can be explained as the joint effect of two processes, namely *social influence* [59] and *selection* [127]. Each user tends to become similar to people he/she interacts with. In the context of modeling users' choices and tastes, the principle of social influence states that the tendency of adopting a new product increases with the number of social peers who have already adopted it. In this sense, social influence promotes homogeneity of tastes/adoptions within the same community. Selection refers to the tendency to create social connection with similar people; it therefore promotes fragmentation of the social network in well-separated, highly homogenous modules. As a result, each user in a network exhibits a high degree of similarity with his/her neighbors.

These effects have been studied in different settings. For instance, in Wikipedia, people tend to edit the same articles after contributing to the same discussion page [46], which is assumed to represent the establishment of their social tie. Furthermore, Aiello et al. [7] found that, within social media systems, users with neighboring relationships exhibit a high level of lexical and topical

---

[8]http://www.lastfm.com/
[9]http://www.flixster.com/
[10]http://www.tripadvisor.com/

alignment: that is, they tend to adopt the same tag, and to access similar groups of resources. Empirical studies [167, 210] have also confirmed the importance of social influence in social recommender systems and empirically measured to what extent users' choices are affected by their social peers.

Despite being relatively new, this field of research is extremely promising, and several adaptations of the probabilistic techniques shown in the previous chapters have been proposed. In addition, new probabilistic models allow us to understand complex phenomena and to exploit them in the recommendation scenario. We shall review these models and their applications in the next sections.

6.1 MODELING SOCIAL RATING NETWORKS

A *Social Rating Network (SRN)* is a platform that allows users to establish social relationships and share explicit opinions and preferences on a set of items. Opinions can be provided in different formats (e.g., positive/negative, explicit ratings, or in textual format), while social links can be either directed (follower/following relationships) or undirected (mutual friendship), and, in the case of *trust networks* [125], can be weighted with a numerical value that expresses the strength of the trust relationship between two peers.

We assume the situation where users provide explicit numerical ratings on items of interest. Then, a SRN can be formalized as a tuple $\langle G = (\mathcal{U}, E), \mathbf{R}, \mathcal{I}, \mathcal{V} \rangle$, where the set of vertices of the social graph G coincides with the user-set $\mathcal{U}$ and $E \subseteq \mathcal{U} \times \mathcal{U}$ encodes the social relationships, while $\mathbf{R}$ stores the numerical preferences, within a fixed scale $\mathcal{V}$, provided by users on the set of items $\mathcal{I}$. Let us introduce a binary friendship indicator $e_{u,v}$, which is equal to 1 if $(u, v) \in E$, zero otherwise. Moreover, let $E^+(u)$ denote the set of users followed by u, i.e., $E^+(u) = \{v \in \mathcal{U} : (u, v) \in E\}$, while symmetrically $E^-(u)$ denotes the set of user following u. A toy example of a directed SRN is given in Figure 6.1. The underlying network includes five users and their preferences expressed on a set of eight items. Both the social network and the preference data are characterized by high sparsity. Table 6.1 lists some publicly available datasets that include both user preferences and social connections.

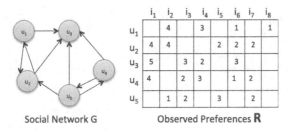

	i_1	i_2	i_3	i_4	i_5	i_6	i_7	i_8
u_1		4		3		1		1
u_2	4	4			2	2	2	
u_3	5		3	2			3	
u_4	4		2	3		1	2	
u_5		1	2		3		2	

Social Network G   Observed Preferences **R**

Figure 6.1: Social rating network.

Table 6.1: Example of publicly available SNR datasets

Name	Domain	Social relations	Users' feedback	URL dataset
Epinions	generic	explicit trust	ratings/textual	www.trustlet.org/wiki/Epinions_dataset
Flixster	movies	mutual friendship	ratings	www.sfu.ca/~sja25/datasets/
Douban	generic	mutual friendship	ratings	http://dl.dropbox.com/u/17517913/Douban.zip
Last.fm	music	mutual friendships	listening counts	www.grouplens.org/node/12

To characterize the behavior of users in online SRNs, we can resort to the analysis carried out in [95]. Actions performed in a SRN context can be classified either as social actions (in particular, for establishing a social connection) or item adoptions, where the latter involves experiencing a product or providing an explicit feedback on an item. Notably, the creation of social links can be modeled as the effect of *transitivity*, i.e., the tendency of establishing social relationships with people at a short distance in the network (friends of friends). Also, adoption actions can be explained in terms of either social influence or other external factors. The evolution and dynamics of a SRN can be analyzed by considering a stochastic model where a new action can be characterized according to three aspects:

- the user performing the action;

- whether the user is performing a social or an adoption action; and

- the influencing factor for the target.

The stochastic model is depicted in Figure 6.2, and is characterized by the parameter set $\Theta = \{\eta, \varphi, \Omega_1, \Omega_2, \Omega_3, \theta_1, \theta_2, \theta_3\}$. Within the model, Ω_1 and Ω_2 expresses the probability of observing a social action due to either transitivity or similarity, respectively. Similarly, θ_1 expresses the probability of observing an adoption action due to social influence, whereas θ_2 is the strength of adoptions based on similarity.

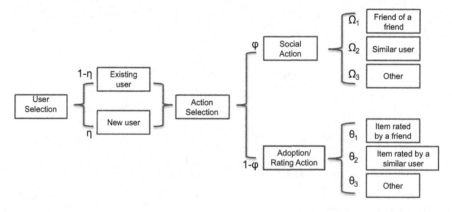

Figure 6.2: Probabilistic model for users' behavior in a SRN.

The study of the parameters Ω and θ allows us to understand how a SRN can evolve, and, in particular, to evaluate the influence of social relationships on the behavior of the users. For example, let $S_t = \{G_t, \mathbf{R}_t\}$ denote the status of the network at time t and the log of all adoptions performed by t. Assuming that, given the state of the SRN, social actions are i.i.d, the likelihood of the social graph can be expressed as:

$$
\begin{aligned}
P(G_t | \Theta_{trans}, \Theta_{ssim}, \Theta_{ext}, \omega_1, \omega_2) &= \prod_{\langle u, v \rangle \in E_t} P(\langle u, v \rangle | t, \Theta_{trans}, \Theta_{ssim}, \Theta_{ext}, \omega_1, \omega_2) \\
&= \prod_{\langle u, v \rangle \in E_t} P(u|t) \{\Omega_1 \cdot P(v|t, \Theta_{trans}) + \Omega_2 \cdot P(v|t, \Theta_{ssim}) \\
&\qquad + (1 - \Omega_1 - \Omega_2) \cdot P(v|t, \Theta_{ext})\},
\end{aligned}
$$

where $P(u|t)$ represents the probability that a new social connection will be established by u at time t, while Θ_{trans}, Θ_{ssim}, and Θ_{ext} are the parameters specifying the probabilities of v being the target of the social connection due to transitivity, similarity, or external factors, respectively. The likelihood for the adoption log $\mathbf{R}_t$ can be expressed similarly. Jamali and Ester in [95] fit the above model by MLE on two real-life datasets, and discover that both the creation of new social connections and the adoption of new items are strongly motivated by social factors ($\Omega_1 \approx 0.9$ and $\theta_1 \approx 0.55$).

6.2   PROBABILISTIC APPROACHES FOR SOCIAL RATING NETWORKS

In the context of a SRN, observations $\mathcal{X}$ can be partitioned into $\mathcal{X}_R$ and $\mathcal{X}_N$, where the former is concerned with adoptions and the latter is concerned with social connections. Most of the techniques developed for rating prediction, and in general for modeling preference matrices, can be applied directly for predicting missing data in network adjacency matrices, as the two tasks are fundamentally the same. The latter setting, known as *link prediction* (for a survey see [9]), can be formalized as follows: given a snapshot of a social network, can we infer which social interactions are more likely to be established in the near future?

Here, we consider $\mathcal{X}_N$ as side information that can help in improving the modeling of users in those scenarios where $\mathcal{X}_R$ is particularly poor in explicit preference observations (e.g., in cold-start scenarios). Approaches to SRNs can be divided roughly into two classes, which correspond to different ways of modeling how the social interactions affect the user behavior. The first class comprises all those methods which *explicitly* model dependency among users' choices. These approaches relax the assumption that users are independent from each other, by modeling each user's choice as a function of the choices performed by her social peers. Two examples are random walks on the friendship network and elicitation of influence among neighbors. The former approaches assume that ratings expressed by trusted friends on similar items can be more reliable than ratings expressed by far neighbors in the social network on the target item: *TrustWalker* [94], for example, employs a random walk over the network to predict the user's preference on a considered item by basing the prediction on the preferences expressed on similar items by social peers at a short distance within the social graph. The latter approaches assume that the tendency of a user to adopt an item is a monotone function of the number of his social peers who have already adopted the same product. We shall consider in detail the study of influence within social network in Section 6.3.

The second class of approaches is based on the idea of the joint modeling of social relationships and users' preference to better define hidden relationships in the user latent space. This is obtained by employing the same set of latent variables and ensures that the projections of two users into the latent space will become closer if they share common friends or exhibit similar preference patterns. An example is given by the *social regularization* [119] approach for matrix factorization, in which the objective function is formulated to include a regularization parameter that penalizes the distance between latent vectors of users who are connected in the network. The remainder of the section is devoted to discuss how to reformulate a network-aware latent factor modeling in a probabilistic setting.

6.2.1   NETWORK-AWARE TOPIC MODELS

In a topic-modeling scenario, one can think of directly exploiting the idea of regularization by adding further constraints on the user-topic probabilities [128]. That is, the effect of the social connections on a model Θ is expressed by the fact that, for a given pair u and v, the probabilities θ_u and θ_v must be similar when u and v are socially connected. The smoothing of the resulting topic model can be obtained by considering the joint likelihood

$$P(\mathcal{X}_R, \Theta | G) = P(\mathcal{X}_R | \Theta) P(\Theta | G), \tag{6.1}$$

where the posterior $P(\Theta | G)$ expresses the above constraints, e.g.,

$$P(\Theta | G) \propto \prod_{\langle u, v \rangle \in E} \exp \left\{ \left(\theta_{u,k} - \theta_{v,k} \right)^T \Sigma_{u,v}^{-1} \left(\theta_{u,k} - \theta_{v,k} \right) \right\}.$$

An alternative approach is to consider both item adoptions and social connections as the result of a stochastic process governed by latent factor. For example, we can extend the LDA to embody a probabilistic model for social links [55]. Figure 6.2.1 shows an example of such a model, where the generation of links follows a process similar to the generation of items and the user-specific distribution θ_u is used to sample latent topics for both adopted products and social connections.

The generative model is described as follows.

1. For each latent factor $k = 1, \ldots, K$ sample:

   - a multinomial distribution over items $\phi_k \sim Dir(\beta)$;

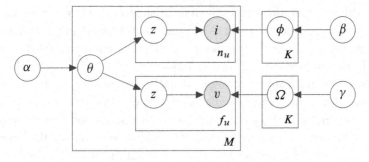

Figure 6.3: Graphical model for Link-LDA. Here $f_u = |E^+(u)|$.

- a multinomial distribution over users $\Omega_k \sim Dir(\gamma)$.

2. For each user $u \in \mathcal{U}$:

   (a) Choose $\theta_u \sim Dir(\alpha)$;

   (b) Sample the number n_u of item selections;

   (c) For each of the n_u items to be generated:

      i. Sample a topic $z \sim Disc(\theta_u)$;

      ii. Sample $i \sim Disc(\phi_z)$;

   (d) Sample the number f_u of social links;

   (e) For each of the f_u social links to be generated:

      i. Sample a topic $z \sim Disc(\theta_u)$;

      ii. Sample a target user for the social link $v \sim Disc(\Omega_z)$.

Here, the state z of the latent variable identifies both an abstract pattern of adoption and a social community.

As usual, the joint likelihood can be expressed as:

$$P(\mathcal{X}|\alpha, \beta, \gamma) = P(\mathcal{X}_R|\alpha, \beta) \cdot P(\mathcal{X}_N|\alpha, \gamma)$$

$$= \int \int \int \prod_u \left[\prod_{i \in \mathcal{I}(u)} \sum_k \theta_{u,k} \, \phi_{k,i} \right] \cdot \left[\prod_{v \in E^+(u)} \sum_k \theta_{u,k} \, \Omega_{k,v} \right] \cdot \qquad (6.2)$$
$$P(\Theta|\alpha) P(\phi|\beta) P(\Omega|\omega) \; d\Theta \, d\Phi \, d\Omega$$

The shared latent variable ensures that users who share the same social connections and tend to adopt the same items, will exhibit similar distributions over topics. It is noteworthy to compare

Equation 6.2 with the joint likelihood shown in Equation 6.1. Here, we assume that both $\mathcal{X}_N$ is the result of a stochastic process, where as in Equation 6.1 we assume that the network is fixed.

Also, note that in this generative model social links are sampled from a multinomial distribution which is defined over the set of users. A more precise formulation would replace the multinomial distribution with a *multivariate hypergeometric distribution*, which simulates the sampling, without replacement, from a finite population whose elements can be classified in a set of disjoint classes. However, the multivariate hypergeometric distribution converges to the multinomial when the size of the sampling population is large. As discussed in [42], the large number of users in online SRNs makes the difference between the two distributions negligible, thus simplifying the modeling.

6.2.2 SOCIAL PROBABILISTIC MATRIX FACTORIZATION

In this section, we reformulate the probabilistic modeling of adoptions and social connections by exploiting matrix factorization techniques. For each pair $\langle u, i \rangle$ of observations, a matrix factorization model assumes the existence of latent factors $\mathbf{P}_u$ and $\mathbf{Q}_i$ such that $r_i^u \sim \mathcal{N}\left(P_u^T \mathbf{Q}_i, \Sigma^{-1}\right)$. However, the social connections in G can be expressed by an incidence matrix $\mathbf{E}$ such that $e_{u,v} = 1$ if and only if $(u, v) \in E$. As a result, we can provide a matrix factorization interpretation for this matrix as well. Moreover, since users are already projected into a latent factor space by means of a matrix $\mathbf{P}$, the same factor space can be exploited in the factorization of $\mathbf{E}$.

The *So-Rec* factorization model [118], given graphically in Figure 6.4, assumes that the generation of the network structure and the preference observations are governed by a shared set of latent factors. This joint modeling promotes the projection into the same low-rank latent feature space of users that exhibit similar social connections and/or rating behavior. The model is specified by three sets of latent factors: $\mathbf{P} \in \mathbb{R}^{K \times M}$ represents the interest of each user in following other people or rating products; $\mathbf{Q} \in \mathbb{R}^{N \times K}$ encodes relationships between items and topics; and, finally, $\mathbf{F} \in \mathbb{R}^{K \times M}$ represents the authoritativeness of each user (chance of being followed) in each topic. Given latent factor vectors, the likelihood of the set $\mathcal{X}$ of observations can be expressed as:

$$P(\mathcal{X}|\mathbf{P}, \mathbf{Q}, \mathbf{F}) = \prod_{\langle u,i,r \rangle \in \mathcal{X}_R} P(r|u, i, \mathbf{P}, \mathbf{Q}) \prod_{\langle u,v \rangle \in \mathcal{X}_N} P(e_{u,v}|u, v, \mathbf{P}, \mathbf{F}),$$

where $P(r|u, i, \mathbf{P}, \mathbf{Q})$ and $P(e_{u,v} = 1|u, v, \mathbf{P}, \mathbf{F})$ are instantiated as Gaussian/logistic density functions that have been studied in the previous chapters. Plugging the above data likelihood into the probabilistic matrix factorization framework allows us to solve the related inference and prediction problems, e.g., by means of Gibbs sampling or variational inference.

A particular instantiation of the above models is the case where we enforce $\mathbf{F} = \mathbf{P}^T$. Users can be explained in terms of a unique mapping within the latent factor space, and the incidence matrix $\mathbf{E}$ is explained in terms of the factorization $\mathbf{P}^T \mathbf{P}$. A generic entry $\mathbf{P}_u^T \mathbf{P}_v$ of the latter matrix can be interpreted as the cosine similarity of users within the latent factor space. Hence,

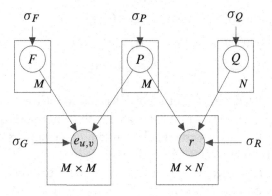

Figure 6.4: Graphical model for *So-Rec*.

the factorization of **E** provides an explanation of symmetric networks in terms of similarity: two users are likely to exhibit a connection if they share the same interests in the latent factor space.

6.2.3   STOCHASTIC BLOCK MODELS FOR SOCIAL RATING NETWORKS

We can generalize the joint factorization of both the social graph and the users' preferences by resorting to *mixed membership stochastic blockmodels (MMSB)* [8, 96]. Again, we assume that a generic user can be associated with a topic distribution θ_u. Furthermore, we assume that items can be associated with a topic distribution as well. Topic distributions influence item adoptions and social connections. However, similar to the co-clustering models shown in Chapter 3, there are two "roles" involved in each action, where each role refers to a specific topic. The generative process of the *Social-aware stochastic block model* summarizing these concepts is shown below.

1. For each user $u \in \mathcal{U}$, sample the mixed membership profile $\theta_u \sim Dir(\alpha^1)$.

2. For each user $i \in \mathcal{I}$, sample the item membership vector $\psi_i \sim Dir(\alpha^2)$.

3. For each pair $(z, w) \in \{1, \ldots K\}^2$ sample $\beta_{z,w} \sim Beta(a, b)$.

4. For each pair $(z, w) \in \{1, \ldots K\} \times \{1, \ldots, L\}$ sample $\epsilon_{z,w} \sim Dir(\gamma)$.

5. For each pair of user $(u, v) \in \mathcal{U} \times \mathcal{U}$:

   • Sample a group for the source node $z_l \sim Disc(\theta_u)$;

   • Sample a group for the destination node $w_l \sim Disc(\theta_v)$;

   • Generate the social relationship $e_{u,v} \sim Bernoulli(\beta_{z_l, w_l})$.

6. For each user $u \in \mathcal{U}$, and for each item $i \in \mathcal{I}(u)$:

- Sample a group for the user $z^a \sim Disc(\boldsymbol{\theta}_u)$;
- Sample an item category for the item $w^a \sim Disc(\boldsymbol{\Phi}_i)$;
- Generate a rating value $r_i^u \sim Disc(\epsilon_{z^a, w^a})$.

Each user is associated with a multinomial distribution that models her membership over a finite set of K groups, while each item is associated with a multinomial distribution over a set of L categories. The probability of observing a link among the members of two groups is governed by a $K \times K$ matrix of Bernoulli rates. For each possible social link (u, v), we sample: (i) a group membership for the source node; (ii) a group membership for the target node; and, (iii) finally we generate a link according to the Bernoulli distribution β, which encodes the interactions among the groups. Analogously, a preference is generated by choosing the topics z^a and w^a associated to each pair $\langle u, i \rangle$; then we sample a rating by drawing upon the multinomial distribution that governs preferences for users belonging to the group z^a, on items belonging to the category w^a.

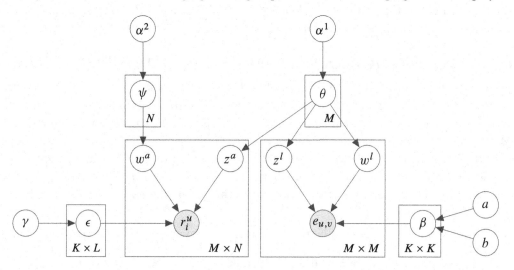

Figure 6.5: Social-aware stochastic block model.

The joint likelihood is quite straightforward: by assuming that $\mathbf{Z}^a$, $\mathbf{Z}^l$, $\mathbf{W}^a$, and $\mathbf{W}^l$ represent the matrices of topic assignments,

$$P(\mathcal{X}, \mathbf{Z}^a, \mathbf{Z}^l, \mathbf{W}^a, \mathbf{W}^l | \beta, \theta, \psi, \epsilon) = P(\mathcal{X}_N, \mathbf{Z}^l, \mathbf{W}_l | \beta, \theta) \cdot P(\mathcal{X}_R, \mathbf{Z}^a, \mathbf{W}^a | \theta, \psi, \epsilon),$$

where, in particular,

$$P(\mathcal{X}_N, \mathbf{Z}^l, \mathbf{W}^l | \beta, \theta) = \prod_{(u,v) \in \mathcal{U} \times \mathcal{U}} P(e_{u,v} | z_{u,v}^l, w_{u,v}^l, \boldsymbol{\beta}) \cdot P(z_{u,v}^l | \theta_u) \cdot P(w_{u,v}^l | \theta_v), \qquad (6.3)$$

and

$$P(e_{u,v}|k,l,\beta) = \cdot\beta_{k,l}^{e_{u,v}} \cdot (1 - \beta_{k,l})^{1-e_{u,v}}.$$

The estimation of the parameters can be performed by either applying variational inference, or by adapting the Gibbs sampling scheme illustrated in Chapter 3. Notice that the likelihood in Equation 6.3 requires us to take into account the whole incidence matrix. As discussed in [68], sampling techniques should be adopted to address the high computational burden of the learning phase, which is strongly influenced by the structure of this component.

6.3   INFLUENCE IN SOCIAL NETWORKS

Understanding the dynamics characterizing social influence and the influence-driven diffusion of information, also known as *information cascades*, in social networks has received growing attention by both academic and industrial communities. The motivating idea is nicely expressed in the seminal work by Gladwell [66]:

> *Ideas and products and messages and behaviors spread like viruses do.*

This motivates a wide range of applications for social influence analysis: viral marketing, identifying influencers and promoting their deeper engagement with the system, generating personalized recommendations based on those influencers, and feed ranking, just to mention a few.

We are still interested in studying the social network $\langle G = (\mathcal{U}, E), \mathcal{I} \rangle$ discussed before, where $(u, v) \in E$ represents the fact that information can propagate from u to v, e.g., u can influence v and $\mathcal{I}$ represents the items that can be adopted and that can spread along the network. The set $\mathcal{X}$ of observations, however, now includes the cases where an individual performs a certain action for the first time: that is, triplets $\langle u, i, t \rangle$ denoting the fact that user u adopted (purchased/rated/clicked) the item i at time t. We can also view the set of observations under a different perspective, i.e., as a set of observed propagation traces $\{\alpha_1, \cdots, \alpha_N\}$ over $\mathcal{I}$, where each trace α contains the activation time of each node $t_\alpha(v)$, where $t_\alpha(v) = \infty$ if v does not become active in trace α. Thus, α induces a directed acyclic graph, where the parents of a node u are

$$par_\alpha(u) = \{v : (v, u) \in E, t_\alpha(v) < t_\alpha(u)\}.$$

The setting described so far is the most general one, where we only observe users adopting some products, without knowing who triggered the adoption, i.e., the actual influencer. For instance, we may record the time at which a user rated a particular item and assume that this item has been propagated from v to u if $(v, u) \in E$ if u rates the item shortly after the rating by v. If we observe this pattern several times, we can infer that v influences u.

Although providing a comprehensive review of the state-of-the-art of social influence analysis is out of the scope of this monograph (the interested reader may refer to [35] for an overview of influence propagation in social networks), in the following, we summarize the main research directions and their relationships to recommender systems.

6.3.1 IDENTIFYING SOCIAL INFLUENCE

Identifying and understanding social influence is a complex task. The simple observation of the high correlation between activations of two social peers is not sufficient for devising social influence. In fact, the statistical correlation among activations of connected peers in a social network can be explained not only by social influence, but also as the product of homophily and/or confounding factors (external factors, such as the environment in which peers are embedded). Figure 6.6 illustrates how these components relate: when both u and v adopt the same item i, there can be three different explanations:

- The environment where these users are located "forces" them to choose the same adoptions (e.g., they have access to a catalog which only makes i available).

- They share common tastes and preferences, which are expressed by similar choices (homophily).

- v recognizes u as an authoritative source of influence, and hence, she tends to emulate u's behavior.

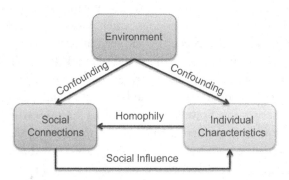

Figure 6.6: Sources of social correlations (adapted from [190]).

Identifying these settings in which social influence plays an important role in shaping users' behavior is important to design viral marketing campaigns that exploit "word of mouth" phenomena and the role of influential users. Distinguishing social influence from other sources of correlation essentially boils down to distinguishing correlation from causality, which is a hard statistical problem. In [11], authors propose two tests for deciding if influence is the source of correlation. The general framework for modeling users' activation is based on a logistic function: the probability of becoming active is a monotone increasing function of the number of social peers already active:

$$P(\text{activation happens when } a \text{ peers are already active}) = \frac{\exp\{\beta_1 \ln(a+1) + \beta_0\}}{1 + \exp\{\beta_1 \ln(a+1) + \beta_0\}},$$

where a is the number of active peers, β_1 measures social correlation, and β_0 is the intercept. Regression coefficients can be estimated by employing maximum likelihood estimation for the logistic regression.

The first test, called *shuffle test*, is based on the following idea: if adoptions are not due to influence, then, while the probability of activation still depends on the number of friends that are already active, the timings of these activations should be independent. More in details, the shuffle test initially computes the value of β_1, which maximizes the likelihood of observing the real data. If we assume that the activation time of each node is independent, identically distributed from a distribution over the observation window, then the maximum likelihood estimation of β_1 is close to its expected value, where the expectation is taken over the random choice of the time stamps. That is, denoting by π a random permutation of the activation timestamps, the shuffle test declares that the model exhibits no social influence if the values of β_1 and $\tilde{\beta}_1$ are close to each other where the latter is the regression coefficient computed by considering the random permutation π.

The *edge-reversal* test is based on the idea that, if correlation is due to homophily or confounding factors, then reversing the direction of edges should not affect the estimate of the social correlation significantly. On the other hand, if influence is the cause of correlation, then reversing the direction of the edges leads to a different estimate of the social correlation parameter β_1.

Both tests are effective in detecting instances in which influence is the source of correlation, and they provide a qualitative indication of the existence of influence. Measuring the extent of influence and handling the case where different source of correlations may simultaneously affect users' behavior however, require more refined modeling.

6.3.2   INFLUENCE MAXIMIZATION AND VIRAL MARKETING

A natural scenario where social influence analysis and recommendation techniques find an intersection is *viral marketing*. The latter refers to marketing techniques aimed at exploiting social media and communication channels to promote products or brands. For example, we might be interested in devising, within the network G and for a specific item i, a minimal set of users to target such that the number of adoptions triggered by peer-wise influence is maximized. This problem was initially tackled by Domingos and Richardson [53, 163], who modeled the diffusion process in terms of Markov random fields.

Later, Kempe et al. [104] proposed a formulation of the setting studied by Domingos and Richardson in terms of a discrete optimization problem, focusing on two fundamental propagation models: the *Independent Cascade* (IC) and *Linear Threshold* (LT) models. Both focus on binary infections: at a given timestamp, each node is either active (a user who already adopted the item), or inactive, and an active node never becomes inactive again. Time unfolds deterministically in discrete steps and each node's tendency to become active increases monotonically as more of its neighbors become active. Under the IC model, each new active node v at time t is considered contagious and has one chance of influencing each inactive neighbor u, independent of the history thus far. The activation of u by v succeeds with Bernoullian probability $p_{v,u}$; the activation trial is unique, as u cannot make further attempts to activate u in subsequent rounds. According to the LT model, each node u is influenced by each active neighbor v according to a multinomial weight $p_{v,u}$, such that the sum of incoming weights to u is no more than 1. Each node u is associated with a "resistance" threshold θ_u. At any timestamp t, if the total weight from the active neighbors of an inactive node u is at least θ_u, then u becomes active at timestamp $t + 1$. In both cases, the process runs until no more activations are possible.

Given a propagation model m (either IC or LT) and a seed set $S \subseteq V$, the *spread* of S, i.e., the expected number of active nodes at the end of the propagation process, is denoted by $\sigma_m(S)$. The *influence maximization problem* is to find the set $S \subseteq \mathcal{U}$, $|S| = k$, such that $\sigma_m(S)$ is maximized. Under both the IC and LT propagation models, the problem was proved to be **NP**-hard [104]. There are however two nice properties of the $\sigma_m(S)$ function:

- *monotonicity*, i.e., $\sigma_m(S) \leq \sigma_m(T)$ whenever $S \subseteq T$, and

- *submodularity*, i.e., $\sigma_m(S \cup \{w\}) - \sigma_m(S) \geq \sigma_m(T \cup \{w\}) - \sigma_m(T)$ whenever $S \subseteq T$.

Under these properties, the simple greedy algorithm that at each iteration greedily extends the set of seeds with the node providing the largest marginal gain, as described in Algorithm 4, produces a solution with a provable approximation guarantee $(1 - 1/e)$ [142].

Algorithm 4 Greedy algorithm for influence maximization.

Require: Network G, seed-set budget k, propagation model σ_m.
Ensure: A seed set S, with $|S| = k$, of users to target.
  1: $S \leftarrow \emptyset$
  2: **while** $|S| < k$ **do**
  3:     $u \leftarrow \arg\max_{w \in V \setminus S} (\sigma_m(S \cup \{w\}) - \sigma_m(S))$
  4:     $S \leftarrow S \cup \{u\}$
  5: **end while**

Computing the spread of a given seed set of nodes is #**P**-hard under both the IC and the LT models. An accurate and more feasible estimation can be achieved by running Monte Carlo simulations; as shown in [104] for any $\phi > 0$, there is a $\delta > 0$ such that by using $(1 + \delta)$-approximate values of the expected spread, we can obtain a $(1 - 1/e - \phi)$-approximation for the influence maximization problem. Unfortunately, simulations are extremely costly on very large real-world social networks and considerable effort has been devoted to develop methods for improving the efficiency of influence maximization [43, 44, 72, 105, 113, 165].

Learning Influence Probabilities

Approaches to influence maximization assume that the influence probabilities for each pair of users are given as input. An accurate estimate of the influence weights is hence important. It is worth noticing that the influence weights are not necessarily related to the degree of the nodes [41], and they require ad-hoc estimation techniques by mining an observed set of information cascades.

As described before, each propagation trace α induces a directed acyclic graph, where each link (v, u) represent the fact that the information potentially propagated from v to u. In practice, the induced graph devises, for each user u, the set $par_\alpha(u)$ of possible activators. To prune episodes of propagations that are unlikely, we can introduce a temporal threshold Δ. Let $F^+_{\alpha,u}$ be the set of u's neighbors that potentially influenced u's activation in the trace α:

$$F^+_{\alpha,u} = \{v \mid (v, u) \in E, 0 \leq t_\alpha(u) - t_\alpha(v) \leq \Delta\}.$$

Also, consider the set $A_{v \to u} = \{\alpha \in \mathcal{X} | v \in F^+_{\alpha,u}\}$ of traces for which v is a potential influencer of u, and the set $A_v = \{\alpha \in \mathcal{X} | t_\alpha(v) < \infty\}$ containing all the adoptions performed by v. Simple estimates of influence probabilities can be computed by relating the number of propagation episodes [71], e.g.:

$$p_{v,u} = \frac{|A_{v \to u}|}{|A_v|},$$

or

$$p_{v,u} = \frac{|A_{v \to u}|}{|A_u \cup A_v|}.$$

In both models, each node w in $F^+_{\alpha,u}$ takes the credit for the activation of u. Specific adjustments can be obtained by adopting *partial credit* models [71] to quantify the importance of potential influencers in triggering one activation. In particular, if $|F^+_{\alpha,u}| = 1$, the unique potential influencer should take all the credit for triggering u's activation, while if $|F^+_{\alpha,u}| > 1$, then the overall credit should be shared. We can define the credit as follows:

$$credit_{v,u}(\alpha) = \begin{cases} 0 & \text{if } v \notin F^+_{\alpha,u} \\ \frac{1}{|F^+_{\alpha,u}|} & \text{otherwise} \end{cases},$$

and reformulate the estimates for influence probabilities as follows:

$$p_{v,u} = \frac{\sum_{\alpha \in \mathbb{D}} credit_{v,u}(\alpha)}{|A_v|}, \qquad \text{or} \qquad p_{v,u} = \frac{\sum_{\alpha \in \mathbb{D}} credit_{v,u}(\alpha)}{|A_u \cup A_v|}.$$

The estimation can also be formalized as a maximum likelihood problem [166], and solved by exploiting the EM algorithm. Within the IC model, recall that multiple nodes may succeed to activate, independently, the same node at the same time, and the activation attempt is unique. Denote by $F^-_{\alpha,u}$ the set of u's neighbors who definitely failed in influencing u on α:

$$F^-_{\alpha,u} = \{v \mid (v, u) \in E, t_\alpha(u) - t_\alpha(v) > \Delta\},$$

and let $A_{v \nrightarrow u}$ be the set of propagation traces for which v failed to influence u, i.e., $A_{v \nrightarrow u} = \{\alpha \in \mathcal{X} | v \in F^-_{\alpha,u}\}$. Assuming i.i.d. propagation traces, the log-likelihood of the traces in $\mathcal{X}$, given the parameter set Θ (representing all the possible influence probabilities $p_{u,v}$), can be expressed as

$$\log L(\mathcal{X} \mid \Theta) = \sum_{\alpha \in \mathbb{D}} \log L_\alpha(\Theta),$$

where the likelihood of a single trace α is

$$L_\alpha(\Theta) = \prod_{u \in \mathcal{U}} \left[1 - \prod_{v \in F_{\alpha,u}^+} (1 - p_{v,u}) \right] \cdot \left[\prod_{v \in F_{\alpha,u}^-} (1 - p_{v,u}) \right]. \tag{6.4}$$

The first term of Equation 6.4 represents the fact that, if u is active, then at least one of the neighbors in $F_{\alpha,u}^+$ succeeded in triggering his activation. Dually, the second term models the fact that none of the neighbors in $F_{\alpha,u}^-$ succeeded in activating u.

Let $z_{\alpha;v,u}$ be the binary latent variable denoting the fact that v triggered the activation of u in α, and let $\mathbf{Z}_\alpha$ be the set of all possible latent variables relative to α. We can reformulate Equation 6.4 in terms of joint likelihood:

$$P(\alpha, \mathbf{Z}_\alpha | \Theta) = \prod_{u \in \mathcal{U}} \left[\prod_{v \in F_{\alpha,u}^+} p_{v,u}^{z_{\alpha;v,u}} \cdot (1 - p_{v,u})^{1 - z_{\alpha;v,u}} \right] \cdot \left[\prod_{v \in F_{\alpha,u}^-} (1 - p_{v,u}) \right],$$

and $P(\mathcal{X}, \mathbf{Z} | \Theta) = \prod_{\alpha \in \mathcal{X}} P(\alpha, \mathbf{Z}_\alpha | \Theta)$. Within the standard EM framework, the joint likelihood induces the expectation log-likelihood

$$Q(\Theta; \tilde{\Theta}) = \sum_{\alpha \in \mathbb{D}} \sum_{u \in V} \left\{ \sum_{v \in F_{\alpha,u}^+} \left(\varphi_{\alpha,v,u} \log p_{v,u} + (1 - \varphi_{\alpha,v,u}) \log(1 - p_{v,u}) \right) + \right.$$
$$\left. \sum_{v \in F_{\alpha,u}^-} \log(1 - p_{v,u}) \right\}, \tag{6.5}$$

where $\varphi_{\alpha,v,u}$ represents the probability that, in trace α, the activation of u was due to the success of the activation trial performed by v. The optimization is achieved by alternating the following steps until convergence:

E-Step: Estimate responsibilities as

$$\varphi_{\alpha,v,u} = \frac{p_{v,u}}{1 - \prod_{w \in F_{\alpha,u}^+} (1 - p_{w,u})};$$

M-Step: Estimate influence probabilities as

$$p_{v,u} = \frac{1}{|A_{v \to u}| + |A_{v \not\to u}|} \sum_{\alpha \in A_{v \to u}} \varphi_{\alpha,v,u}.$$

Interesting variations of the above modeling consider the inclusion of latent topics in the exertion of influence. In real-world scenarios, we naturally trust some people on some topics and others on different topics. Users authoritativeness, expertise, trust, and influence are clearly topic-dependent and the characteristics of the item being the subject of the viral marketing campaign should be considered to design better influence propagation models [117, 139].

6.3.3 EXPLOITING INFLUENCE IN RECOMMENDER SYSTEMS

How to exploit influence and "word-of-mouth" propagations in the context of RSs is still an open research problem. The identification of the influential peers for the user corresponding to the current browsing session enables more effective social explanations [183] in the form "Your friend X likes this product." Alternatively, one could provide incentives to influential people if they proactively provide explicit suggestions of a product to their peers, in the form of "Suggest to friends".

In a SRN setting, both positive and negative opinions expressed by social peers play a role in promoting or lowering the likelihood of adopting an item. Propagation models can be extended easily to cope with such situations, as shown, e.g., in [182]. In practice, each user can have three possible states relative to an item i, namely like, dislike, or stay inactive. The activation in this case is either positive or negative, depending on the actual user's preference on the considered item. Influence can be exerted either positively or negatively: for example, in the LT model, a user's preference becomes negative if the weights of all her potential influencers who disliked the product overcomes the weights of those neighbors who liked it, and, in general, the sum of such weights is above a given threshold. Given an initial state, and the model parameters, we can study the evolution of preferences, and then apply the result to predict the preferences of each user in the network.

More generally, item adoptions can be expressed as a direct effect of influence probabilities,

$$P(i|u) = \sum_{v \in E^-(u)} P(i|v) p_{v,u},$$

thus relating the probabilities of adoptions to the eigenvectors of $\mathbf{P}$. Starting from this simple model, other more sophisticated models can be devised: for example, we can express the probability of adoption in a topic-wise fashion [203] according to the following generative process, shown graphically in Figure 6.7:

1. For each preference observation to be generated:

   (a) sample a user $u \sim Disc(\mathbf{\Pi})$;

   (b) sample an influencer $f \sim P(\psi_u)$;

   (c) sample a topic $z \sim Disc(\theta_f)$;

   (d) sample $i \sim Disc(\phi_z)$.

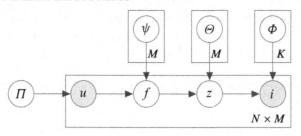

Figure 6.7: Graphical model for social-influence-based recommendations.

Here we assume that each user is associated with a multinomial distribution ψ_u over possible influencers; the probability of adoption of a particular item depends on the topic distribution of the previously chosen influencer:

$$
\begin{aligned}
P(i|u; \Psi, \Theta, \Phi, \Pi) &= \frac{P(i, u|\Psi, \Theta, \Phi, \Pi)}{P(u|\Pi)} \\
&= \frac{\sum_z \sum_f p(i, u, f, z|\Psi, \Theta, \Phi, \Pi)}{P(u|\Pi)} \\
&= \frac{P(u|\Pi) \sum_z \sum_f p(f|\psi_u) P(z|\theta_f) P(i|\phi_z)}{P(u|\Pi)} \\
&= \sum_z \sum_f p(f|\psi_u) P(z|\theta_f) P(i|\phi_z).
\end{aligned}
$$

The connections between recommendation and influence also can be studied under an influence maximization perspective [73]. Given a generic RS algorithm, we can reformulate the influence maximization problem as the problem of identifying the seed set of influential users, who, by providing high rating to a product, guarantee a maximum number of recommendations. This problem has proven to be NP-Hard and the optimal solution cannot be approximated within any reasonable factor. Goyal et al. in [73] propose several heuristics that are reported to work well in practice on both user and item-based RSs.

CHAPTER 7

Conclusions

There is a recurring persuasion underlying the research covered in this book. Probabilistic methods provide a robust and mathematically elegant tool for modeling preference data, and have revealed extreme flexibility to accommodate different situations. It is worth clarifying that the main thesis here is not the general superiority of probabilistic methods. It is well-known, e.g., from the Netflix prize, that the best approaches count an ensemble of methods that cooperate for a best prediction. Nevertheless, as also witnessed by the studies discussed in this book, probabilistic methods can play a prominent role within such ensembles.

Probabilistic approaches offer some advantages over other techniques, both in terms of modeling and accuracy of the results. We have discussed, in Chapter 2, the capability to differentiate between *free* and *forced* prediction. Also, if we interpret the recommendation as missing value prediction (matrix completion, where we are asked to estimate numerical users preference values on unseen items), the use of Bayesian inference techniques, discussed in Chapter 3, is better suited to handle the sparsity of preference data. Notably, techniques based on Bayesian matrix factorization achieve the highest accuracy in rating prediction, as illustrated in Chapter 4.

Prediction accuracy is commonly used as a proxy metric to measure the effectiveness of the RS. However, recent studies have pointed out many limitations of this approach, showing that higher prediction accuracy does not imply higher recommendation accuracy. We investigated this aspect in Section 4.1.2: when measuring recommendation accuracy as precision and recall of the recommendation lists provided to users, probabilistic models, equipped with the proper ranking function, exhibit competitive advantages over state-of-the-art RSs. In particular, strategies based on item selection guarantee high accuracy values, which can be further boosted by integrating a relevance-ranking function estimating the probability that a user will play and like a given item. This approach combines the benefits of the modeling of preference values (explicit feedback) and of the implicit users selection of items (implicit feedback).

Besides rating prediction and recommendation accuracy, there are other significant advantages in the adoption of probabilistic models for modeling preference data. A successful recommendation should answer to the simple question "What is the user actually looking for?" which is strictly tied with dynamic user profiling. Understanding user preferences enables better decision making and targeted marketing campaigns. In this direction, probabilistic approaches enable the discovery of topics and abstract preference patterns, as well as the identification of homogenous groups of similar-minded people or similar products. The high-dimensional preference data exhibits both global and local patterns, which can be identified and characterized by employing probabilistic co-clustering approaches. In general, co-clustering techniques are better suited to model the mutual relationship between users and items: similar users are detected by taking into account their ratings on similar items, which in turn are identified considering the ratings assigned by similar users.

These scenarios can be better illustrated by the schema in Figure 7.1. Each scenario here is related to specific application requirements. For example, accurate rating prediction is crucial in situations where the recommendation has to effectively match the preferences of users. An example in this setting is, e.g., a peer-review system, where the explicit preference represents the confidence of the reviewer on the subject relative to the item to review. In this scenario, it is important to suggest, to the user, items that she is able to review with confidence. By contrast, the recommendation accuracy is a fundamental tool when the objective is to recommend products that will likely capture the interest of a user, somewhat independently from the explicit preference of the user. Finally, pattern discovery is aimed at a better knowledge of the customer base and/or item catalog, whereas serendipity is aimed at implementing specific marketing strategies, which may also depend on external factors (such as the objective to promote some products over others).

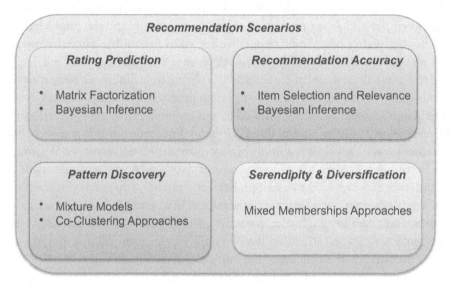

Figure 7.1: A schema of recommendation scenarios and the relative best-suited probabilistic approaches.

Another significant advantage of probabilistic modeling is the capability to accommodate both collaborative and content features in a unified mathematical framework, as we summarize in Chapter 6. This is expected to increase the accuracy of the recommendations provided by the system, and the background content information can be used to provide personalized recommendations in cold-start scenarios. Side information can be exploited for a better identification and interpretation of user communities and item categories, and can be used for a more accurate modeling of the preference data.

A specific context is represented by social connections. Users' behavior on the Web is more and more influenced by their social interactions with other users, and *social recommender systems*, reviewed in Chapter 7, are emerging as a powerful combination of both recommendation and social networking features. In these systems, people *share* content and information, *interact* with their social peers, and *express* evaluations of them, typically in the form of thumbs up/down. This implies a radical change in the recommendation perspective: while in traditional personalized approaches we model users' selections as a process that depends on her past purchase history, here we have to model users' choices as *social process*. Once again, stochastic processes based on latent factors provide an accurate modeling capable of accommodating all these features in a unified framework.

In an attempt to provide a unified picture of all the aspects covered in this book, Figure 7.2 provides a taxonomy of the approaches discussed so far. We can see five (possibly overlapping) categories where each approach fits, and a set of features characterizing each approach. Although the survey provided in this book is not exhaustive, it has the ambition of being paradigmatic, as it is likely to fit any approach not covered here in the scheme provided by the figure, and hence, to serve for the application scenarios illustrated in Figure 7.1.

We would like to conclude this book by presenting some open challenges that probabilistic approaches to recommendation still pose, and whose solutions are the key to a successful affirmation of these approaches in an industrial setting.

7.1 APPLICATION-SPECIFIC CHALLENGES

Although desirable, it is quite difficult to translate theoretical results into practice, and recommender systems are no exception. In Chapter 1, we discussed some metrics for measuring the effectiveness of a recommender system, and, in Chapter 4, we have shown empirical results that demonstrate how probabilistic methods can be tuned to provide optimal results for such metrics. However, it is still an open question how industries can actually leverage the results of this research and successfully deploy the techniques in a real-world scenario. We have discussed accuracy, precision, recall, and RMSE. These measures are widely known by the machine learning researchers and data-mining specialists. However, it is also important to develop economics-oriented measures that capture the business value of recommendations, such as return on investments (ROI) and customer lifetime value (LTV) measures. Notably, companies are looking for ways to increase their sales. Recommender systems are effective under this perspective when:

- customers spend less time searching for products;

- customer satisfaction is increased;

- customer loyalty is increased;

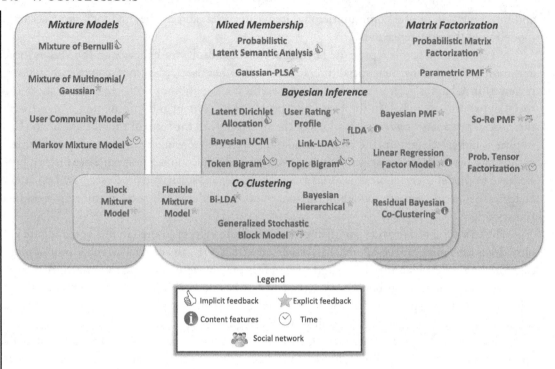

Figure 7.2: A taxonomy of probabilistic approaches to collaborative filtering.

- cross-selling is increased.

The gain of a recommender system depends on the expected increase in cross-selling, the expected likelihood that an increase in customer satisfaction corresponds to an increase in purchases, and the increase of customer loyalty. In order to evaluate such a gain, we need accurate methods to map machine-learning specific metrics to the above expectations. Developing and studying these aspects can help to better evaluate the impacts of the results illustrated in Chapter 4 on industrial scenarios.

We can attempt an academic exercise on the results shown in Figure 4.8. Specifically, we can (arbitrarily) assume that 10% of what is suggested in a recommendation list represents items that a user would not have adopted otherwise. Hence, we can map the recall of a recommender system directly to the expected increase in sales that it would produce. In particular, by comparing the accuracies of PMF and LDA in Figures 4.7 and 4.8, we can conclude that the latter would produce an increase of 4.4% over a 2.5% of the former, on a recommendation list of size 10. However, as already mentioned, the above assumptions are definitely arbitrary, as we do not really know the likelihood that a recommended product wouldn't be adopted by the user anyway. Other studies [26, 40] propose some alternate protocols, but the whole issue still needs a more rigorous mathematical treatment.

Besides the above mentioned challenge, there is a sort of reversed perspective that is challenging as well. We have argued in Chapter 1 that there are other metrics beyond prediction accuracy that should be considered. Next, we have shown in Chapter 4 that probabilistic methods based on topic modeling can be profitably exploited in scenarios where we are interested in capturing concepts such as novelty, serendipity, and diversification. Notably, probabilistic graphical models provide several components which can be exploited fruitfully in focusing on the usefulness of the recommendation [126]. However, the objective measurement of concepts such as non-triviality, serendipity, user-needs, and expectations is still an open issue. Although some initial attempts have been made in this respect [61, 111, 137, 193, 194], we still do not know how and in what respect the methods proposed in Chapter 4 are effective, and whether they can be improved further by focusing on adequate loss functions.

Finally, an aspect we did not cover in this book is the relationship between recommendation and topical and geographical diversity. The advent of new technologies includes the user in a mobile context, where a user can express different expectations and needs depending on the geographical location. The capability to track a user through mobile devices represents an opportunity for the recommender system to better tune the recommendations to the user needs, even by taking into account the Markovian nature of a user's location. Combining the probabilistic framework for preference modeling with the ones proposed for geo-location can really project [87] recommendation systems to a new dimension, where the geo-location context is better modeled.

7.2   TECHNOLOGICAL CHALLENGES

From a technological point of view, the aspect *scalability* of probabilistic methods deserves a more deep analysis. The empirical study in Chapter 4 clearly denotes how the performance of some methods, in terms of learning time, especially when Bayesian modeling is taken into account, are particularly problematic.

This issue involves both the volume of preference data and the number of latent factors employed in the learning phase. We would therefore like to have the tools to build latent factor models that scale to large values. Traditional inference techniques, such as Gibbs sampling and variational inference, do not scale easily to huge volume of data. In such cases it is very time-consuming to run even a single iteration of the standard collapsed Gibbs sampling or variational Bayesian inference algorithms. In such cases, it is mandatory to reformulate the learning algorithms in terms of distributed processing based on scalable architectures.

Besides scalability, there are some issues concerned with the fitting procedures. In fact, the methods shown in this book do not account for incremental maintenance. As long as new data is collected, the beliefs embodied in the models need to be revised, and the only revision possible is to learn the model again. Again, the current literature is focusing on these problems under different perspectives: *incremental* and *online learning* [10, 15, 92, 202], *topic evolution,* and *dynamics* [67, 164].

A further question that can be posed is how robust are the proposed methods of attack. Robustness refers to the ability of a system to operate under stressful conditions [90]. There are several stresses that can be applied to a recommender system [136], the most significant being the *dataset stress*. This typically happens when the preference data is particularly noisy or erroneous. In addition, preference data may be corrupted on purpose, for example by injecting ad-hoc profiles to influence the overall performance of the system by pushing or nuking a product. Collaborative recommender systems are extremely vulnerable to attacks that seek to manipulate recommendations made for target items. Hence, it is natural to investigate the vulnerability of probabilistic methods, and to study countermeasures.

Finally, all latent factor models discussed in this book require the number of clusters, topics, or, more generally, latent dimensions, to be specified *a priori.* The choice of the number of latent factors is a delicate matter, as by increasing this number, we typically increase the quality of the fitting at the cost of a higher learning time and increasing the risk of overfitting. We have discussed this aspect in Section 3.1, and have shown some solutions. Recently, Bayesian nonparametric approaches, such as the Chinese restaurant or Indian buffet process (see [64] for a survey), address this issue in a more general way, by automatically estimating how many latent dimensions are needed to fit the observed data, and by allowing the allocation of new clusters as new data is observed. Recent works [201, 204] have shown that the application of nonparametric methods to the recommendation scenario is both effective and practical on large-scale datasets.

APPENDIX A

Parameter Estimation and Inference

Let us consider a set $\mathcal{X} = \{\mathbf{x}_1, \ldots, \mathbf{x}_N\}$ of observations and a general latent parameter set $\mathbf{Z}$, so that the likelihood $P(\mathcal{X}|\Theta)$ can be expressed in terms of the joint distribution $P(\mathcal{X}, \mathbf{Z}|\Theta)$:

$$P(\mathcal{X}|\Theta) = \int P(\mathcal{X}, \mathbf{Z}|\Theta)\, d\mathbf{Z}. \tag{A.1}$$

This appendix covers some mathematical tools for approaching both the parameter estimation and the inference problems within such a framework. Most of the methods covered in this book represent instantiations of the algorithms covered in the following, and the general approaches introduced here can be considered as a prerequisite for understanding such methods.

A.1 THE EXPECTATION MAXIMIZATION ALGORITHM

The *Expectation-Maximization* (EM) algorithm is a general method for deriving maximum likelihood parameter estimates from incomplete (i.e. partially unobserved) data. Within Equation A.1, we are interested in estimating the optimal Θ that maximizes $P(\mathcal{X}|\Theta)$. However, the likelihood is expressed in terms of the components $P(\mathcal{X}, \mathbf{Z}|\Theta)$, which in turn depend on the latent (unknown) variable $\mathbf{Z}$. In this setting, the EM algorithm defines an iterative procedure for estimating Θ.

A typical example is given when the latent space $\mathbf{Z}$ is defined over discrete latent factors $z_{i,k} \in \{0, 1\}$, with $i = 1, \ldots, N; k = 1, \ldots, K$, such that the probability of observing $\mathbf{x}_i$ depends on a single latent factor. That is, $\sum_k z_{i,k} = 1$ and each latent factor express a parameter space Θ_k which defines $P(\mathbf{x}|\Theta_k)$ and

$$P(\mathbf{x}_i|\mathbf{z}_i, \Theta_1, \ldots, \Theta_K) = \prod_{k=1}^{K} P(\mathbf{x}_i|\Theta_k)^{z_{i,k}}.$$

In this case, the prior probability of observing a latent vector $\mathbf{z}_i$ is given by a multinomial distribution parameterized by $\boldsymbol{\pi} = \{\pi_1, \ldots, \pi_K\}$, that is,

$$P(\mathbf{z}_i|\boldsymbol{\pi}) = \prod_{k=1}^{K} \pi_k^{z_{i,k}}.$$

Thus, denoting $\Theta \triangleq \{\Theta_1, \ldots, \Theta_K, \pi\}$, we can define the join likelihood $P(\mathcal{X}, \mathbf{Z}|\Theta)$ as:

$$P(\mathcal{X}, \mathbf{Z}|\Theta) = P(\mathcal{X}|\mathbf{Z}, \Theta) \cdot P(\mathbf{Z}|\pi)$$

$$= \prod_i \prod_{k=1}^K P(\mathbf{x}_i|\Theta_k)^{z_{i,k}} \cdot \prod_i \prod_{k=1}^K \pi_k^{z_{i,k}},$$

and the data likelihood of Equation A.1 can be simplified into

$$P(\mathcal{X}|\Theta) = \sum_{\mathbf{Z}} P(\mathcal{X}, \mathbf{Z}|\Theta)$$

$$= \sum_{\mathbf{Z}} \prod_i \prod_{k=1}^K P(\mathbf{x}_i|\Theta_k)^{z_{i,k}} \cdot \pi_k^{z_{i,k}} \tag{A.2}$$

$$= \prod_i \sum_{k=1}^K P(\mathbf{x}_i|\Theta_k)\pi_k.$$

Optimizing this likelihood with respect to Θ can be difficult, even for simple functional forms of $P(\mathcal{X}|\mathbf{Z}, \Theta)$. As a consequence, one has to resort to iterative optimization methods in order to estimate the optimal Θ.

One such method is the EM algorithm, which iteratively computes a sequence

$$\Theta^{(0)}, \Theta^{(1)}, \Theta^{(2)}, \ldots, \Theta^{(n)}, \ldots$$

such that, for each t, $P(\mathcal{X}|\Theta^{(t)}) \geq P(\mathcal{X}|\Theta^{(t-1)})$. This approach focuses on the joint likelihood $P(\mathcal{X}, \mathbf{Z}|\Theta)$: If $\mathbf{Z}$ were known, we could rewrite the log-likelihood $\log P(\mathcal{X}, \mathbf{Z}|\Theta)$ in simpler terms. In particular, in the example above we can express the latter as

$$\log P(\mathcal{X}, \mathbf{Z}|\Theta) = \sum_{i=1}^N \sum_{k=1}^K z_{i,k} \{\log \pi_k + \log P(\mathbf{x}_i|\Theta_k)\},$$

and we can see that, for tractable functional forms of $P(\mathbf{x}_i|\Theta_k)$, we can find closed forms for the optimization.

In practice, the latent space $\mathbf{Z}$ is unknown and we can only observe $\mathcal{X}$. However, if the posterior $P(\mathbf{Z}|\mathcal{X}, \Theta)$ is mathematically tractable, we can exploit the *expectation* of the log-likelihood over all the possible latent states $\mathbf{Z}$. If we assume a known prior value Θ' for the parameter set, the expectation of the joint likelihood associated to $\mathbf{Z}$ can be expressed as:

$$\mathcal{Q}(\Theta, \Theta') = \int P(\mathbf{Z}|\mathcal{X}, \Theta') \log P(\mathcal{X}, \mathbf{Z}|\Theta) \, d\mathbf{Z}. \tag{A.3}$$

There is a substantial difference between Equation A.3 and the log-likelihood expressed starting from Equation A.1. The latter exhibits an integral inside a logarithm,

$$\log P(\mathcal{X}|\Theta) = \log \left(\int P(\mathcal{X}, \mathbf{Z}|\Theta) \, d\mathbf{Z} \right),$$

whereas in Equation A.3 we moved the integral outside the logarithm. The importance of the expectation log-likelihood $\mathcal{Q}(\Theta, \Theta')$ lies in two facts: first, $\mathcal{Q}(\Theta, \Theta')$ generally is easier to optimize than $\log P(\mathcal{X}|\Theta)$. Second, a value $\hat{\Theta}$ optimizing $\mathcal{Q}(\Theta, \Theta')$ also improves the log likelihood. In fact, whenever the property

$$\mathcal{Q}(\Theta, \Theta') \geq \mathcal{Q}(\Theta', \Theta')$$

holds, then the related property

$$P(\mathcal{X}|\Theta) \geq P(\mathcal{X}|\Theta')$$

holds as well. To see this, consider the difference $\log P(\mathcal{X}|\Theta) - \log P(\mathcal{X}|\Theta')$. We observe that

$$
\begin{aligned}
\log P(\mathcal{X}|\Theta) - \log P(\mathcal{X}|\Theta') &= \log P(\mathcal{X}|\Theta) \int P(\mathbf{Z}|\mathcal{X}, \Theta') \, d\mathbf{Z} - \log P(\mathcal{X}|\Theta') \int P(\mathbf{Z}|\mathcal{X}, \Theta') \, d\mathbf{Z} \\
&= \int P(\mathbf{Z}|\mathcal{X}, \Theta') \log P(\mathcal{X}|\Theta) \, d\mathbf{Z} - \int P(\mathbf{Z}|\mathcal{X}, \Theta') \log P(\mathcal{X}|\Theta') \, d\mathbf{Z} \\
&= \int P(\mathbf{Z}|\mathcal{X}, \Theta') \left\{ \log P(\mathcal{X}|\Theta) \frac{P(\mathbf{Z}|\mathcal{X}, \Theta)}{P(\mathbf{Z}|\mathcal{X}, \Theta)} \right\} d\mathbf{Z} \\
&\quad - \int P(\mathbf{Z}|\mathcal{X}, \Theta') \left\{ \log P(\mathcal{X}|\Theta') \frac{P(\mathbf{Z}|\mathcal{X}, \Theta')}{P(\mathbf{Z}|\mathcal{X}, \Theta')} \right\} d\mathbf{Z} \\
&= \int P(\mathbf{Z}|\mathcal{X}, \Theta') \log \left\{ \frac{P(\mathbf{Z}, \mathcal{X}|\Theta)}{P(\mathbf{Z}|\mathcal{X}, \Theta)} \right\} d\mathbf{Z} \\
&\quad - \int P(\mathbf{Z}|\mathcal{X}, \Theta') \log \left\{ \frac{P(\mathbf{Z}, \mathcal{X}|\Theta')}{P(\mathbf{Z}|\mathcal{X}, \Theta')} \right\} d\mathbf{Z} \\
&= \int P(\mathbf{Z}|\mathcal{X}, \Theta') \log \left\{ \frac{P(\mathbf{Z}, \mathcal{X}|\Theta)}{P(\mathbf{Z}, \mathcal{X}|\Theta')} \right\} d\mathbf{Z} \\
&\quad - \int P(\mathbf{Z}|\mathcal{X}, \Theta') \log \left\{ \frac{P(\mathbf{Z}|\mathcal{X}, \Theta)}{P(\mathbf{Z}|\mathcal{X}, \Theta')} \right\} d\mathbf{Z} \\
&= \mathcal{Q}(\Theta, \Theta') - \mathcal{Q}(\Theta', \Theta') - \int P(\mathbf{Z}|\mathcal{X}, \Theta') \log \left\{ \frac{P(\mathbf{Z}|\mathcal{X}, \Theta)}{P(\mathbf{Z}|\mathcal{X}, \Theta')} \right\} d\mathbf{Z}.
\end{aligned}
$$

$$(A.4)$$

Consider now the term

$$\int P(\mathbf{Z}|\mathcal{X}, \Theta') \log \left\{ \frac{P(\mathbf{Z}|\mathcal{X}, \Theta)}{P(\mathbf{Z}|\mathcal{X}, \Theta')} \right\} d\mathbf{Z}.$$

Recall Jensen inequality [45] stating that, for a concave function f, the following property holds:

$$E[f(X)] \leq f(E[X]).$$

Since the logarithm is a concave function, we have

$$\int P(\mathbf{Z}|\mathcal{X}, \Theta') \log \left\{ \frac{P(\mathbf{Z}|\mathcal{X}, \Theta)}{P(\mathbf{Z}|\mathcal{X}, \Theta')} \right\} d\mathbf{Z}$$

$$\leq \log \left\{ \int P(\mathbf{Z}|\mathcal{X}, \Theta') \frac{P(\mathbf{Z}|\mathcal{X}, \Theta)}{P(\mathbf{Z}|\mathcal{X}, \Theta')} d\mathbf{Z} \right\} \qquad \text{(A.5)}$$

$$= \log \left\{ \int P(\mathbf{Z}|\mathcal{X}, \Theta) d\mathbf{Z} \right\}$$

$$= 0.$$

Combining the equations A.4 and A.5, we finally obtain

$$\log P(\mathcal{X}|\Theta) - \log P(\mathcal{X}|\Theta') \geq \mathcal{Q}(\Theta, \Theta') - \mathcal{Q}(\Theta', \Theta'). \qquad \text{(A.6)}$$

As a consequence, any improvement on $\mathcal{Q}(\Theta, \Theta')$ represents an improvement on the data likelihood, and we can devise an iterative procedure where, starting from an initial guess $\Theta^{(0)}$, we can proceed with iterative optimization steps over $\mathcal{Q}(\Theta, \Theta^{(t)})$, and the above property ensures that these optimizations reflect on the likelihood as well. The *expectation maximization* optimization procedure alternates two steps.

- Initialize $\Theta^{(0)}$.

- Repeat until convergence, for increasing steps t:

  E Step: Evaluate $P(\mathbf{Z}|\mathcal{X}, \Theta^{(t)})$;

  M step: Evaluate $\Theta^{(t+1)}$ as

  $$\Theta^{(t+1)} = \underset{\Theta}{\operatorname{argmax}} \, \mathcal{Q}(\Theta, \Theta^{(t)}).$$

Typically, the convergence is reached if the difference in log-likelihood is negligible. Also, notice that the M step does not necessarily require an optimal solution. By inequality A.6, it is sufficient to find a value $\hat{\Theta}$ that improves the value of $\mathcal{Q}$. This enables a generalized (albeit slower) approach to the optimization, which is particularly useful in those cases where the optimal solution cannot be found but simple improvements are easier to find in a closed form.

Turning attention back to the example in the beginning, we can see that $\mathcal{Q}(\Theta, \Theta^{(t)})$ can be expressed as:

$$\mathcal{Q}(\Theta, \Theta^{(t)}) = \sum_{i=1}^{N} \sum_{k=1}^{K} \gamma_{i,k}(\Theta^{(t)}) \left\{ \log \pi_k + \log P(\mathbf{x}_i|\Theta_k) \right\}.$$

Where the quantity $\gamma_{i,k}(\Theta^{(t)})$ is called the *responsibility* and represents the posterior probability that observation $\mathbf{x}_i$ is associated with latent factor k:

$$\gamma_{i,k}(\Theta) = P(z_{u,k} = 1|\mathbf{x}, \Theta) = \frac{P(\mathbf{x}|\Theta_k)\pi_k}{\sum_{j=1}^{K} P(\mathbf{x}|\Theta_j)\pi_j}. \tag{A.7}$$

Equation A.7 specifies the E step of the algorithm, which hence depends solely on the value $\Theta^{(t)}$ previously computed and enables the specification of the expectation log-likelihood to optimize.

A.2   VARIATIONAL INFERENCE

As we have seen, a central task in the application of probabilistic models is the evaluation of the posterior distribution $P(\mathbf{Z}|\mathcal{X})$. For many models of practical interest, computing such a posterior distribution is unfeasible: for example, in the case of continuous variables, the required integrations may not have closed-form analytical solutions, while the dimensionality of the space and the complexity of the integrand may prohibit numerical integration. For discrete variables, the marginalizations involve summing over all possible values of $\mathbf{Z}$. Although this is always possible in principle, in practical situations there can be exponentially many hidden states so that an exact calculation is prohibitively expensive.

In such situations, we have to resort to two possible approximation schemes. Stochastic approximations adopt sampling procedures, which we shall discuss in the next section. Here, we provide a brief introduction to *mean field variational inference*, extensively used throughout the manuscript, and which represents an effective deterministic approximation approach. A thorough treatment can be found in [31, 102, 138].

To our purposes, variational approximation can be seen as a generalization of the expectation maximization algorithm described above. Without loss of generality, we can assume $\mathbf{Z}$ continuous and define the likelihood as:

$$P(\mathcal{X}|\Theta) = \int P(\mathcal{X}, \mathbf{Z}|\Theta)\, d\mathbf{Z}.$$

Let us assume that we can express an approximation $q(\mathbf{Z})$ to the posterior distribution $P(\mathbf{Z}|\mathcal{X}, \Theta)$. To our purposes, we assume $q(\mathbf{Z})$ independent of Θ. We can notice the following:

$$\begin{aligned}
\log P(\mathcal{X}|\Theta) &= \log \int P(\mathcal{X}, \mathbf{Z}|\Theta)\, d\mathbf{Z} \\
&= \log \int q(\mathbf{Z}) \frac{P(\mathcal{X}, \mathbf{Z}|\Theta)}{q(\mathbf{Z})}\, d\mathbf{Z} \\
&\geq \int q(\mathbf{Z}) \log P(\mathcal{X}, \mathbf{Z}|\Theta)\, d\mathbf{Z} - \int q(\mathbf{Z}) \log q(\mathbf{Z})\, d\mathbf{Z} \\
&= \mathbb{E}_q[\log P(\mathcal{X}, \mathbf{Z}|\Theta)] + \mathbb{H}[q],
\end{aligned} \tag{A.8}$$

where $\mathbb{E}_q[f(X)]$ represents the expected value of $f(X)$ w.r.t. the distribution q, and $\mathbb{H}[q]$ denotes the entropy of q. Again, the derivation follows from Jensen's inequality and the concavity of the logarithm. We denote the term $\mathbb{E}_q[\log P(\mathcal{X}, \mathbf{Z}|\Theta)] + \mathbb{H}[q]$ by $\mathcal{L}(q, \Theta)$. It is possible to provide a more detailed relationship between the data likelihood and the $\mathcal{L}(q, \Theta)$ term. We recall the Kullback-Leibler divergence [45] between two density functions p and q,

$$\mathbb{KL}(q, p) = -\int q(\mathbf{Z}) \log \frac{p(\mathbf{Z})}{q(\mathbf{Z})} \, d\mathbf{Z},$$

and define $\tilde{P}_{\mathcal{X}, \Theta} \triangleq P(\mathbf{Z}|\mathcal{X}, \Theta)$ Then, it can be easily verified that

$$\log P(\mathcal{X}|\Theta) = \mathcal{L}(q, \Theta) + \mathbb{KL}(q, \tilde{P}_{\mathcal{X}, \Theta}).$$

Notice that Equation A.8 also follows from the fact that $\mathbb{KL}$ is a positive functional.

Thus, for a given parameter set Θ, the term $\mathcal{L}(q, \Theta)$ expresses a lower bound of the likelihood of the data, and, in particular, maximizing $\mathcal{L}(q, \Theta)$ with respect to q is equivalent to minimizing the Kullback-Leibler divergence between $q(\mathbf{Z})$ and the posterior distribution $P(\mathbf{Z}|\mathcal{X}, \Theta)$. Also, when q is optimal, $\mathcal{L}(q, \Theta)$ represents the best possible approximation of $\log P(\mathcal{X}|\Theta)$, hence optimizing $\mathbb{E}_q[\log P(\mathcal{X}, \mathbf{Z}|\Theta)]$ with respect to Θ provides a suitable approximation of the optimal parameter $\hat{\Theta}$ that maximizes $\log P(\mathcal{X}|\Theta)$.

Thus, we can devise a variational EM procedure, where we alternatively optimize $\mathcal{L}(q, \Theta)$:

- Initialize $\Theta^{(0)}$;

- Iterate until convergence, for increasing steps t:

  (E step) Given $\Theta^{(t)}$, compute $q^{(t)} = \text{argmax}_q \, \mathcal{L}(q, \Theta^{(t)})$;

  (M Step) Given $q^{(t)}$, compute $\Theta^{(t+1)} = \text{argmax}_\Theta \, \mathbb{E}_q[\log P(\mathcal{X}, \mathbf{Z}|\Theta)]$.

There are several similarities with the above procedure and the EM procedure discussed in the previous section. In fact, when the posterior is tractable, the optimal value q is given by imposing $q(\mathbf{Z}) \triangleq P(\mathbf{Z}|\mathcal{X}, \Theta)$. In this case, the above procedure coincides with the standard EM procedure outlined before. The problem arises when $P(\mathbf{Z}|\mathcal{X}, \Theta)$ is not tractable. In that case, q represents an approximation that can be mathematically tractable. In particular, we can assume a functional form $q(\mathbf{Z}) \equiv q(\mathbf{Z}|\Lambda)$, which depends on some variational parameters Λ, and optimize $\mathcal{L}(q, \Theta^{(t)})$ with respect to such parameters. Exact equations for such optimizations can be obtained when $\mathbf{Z}$ can be decomposed into $\mathbf{Z}_1, \dots, \mathbf{Z}_n$, and we can provide a factorized definition for q:

$$q(\mathbf{Z}|\Lambda) = \prod_i q_i(\mathbf{Z}_i|\Lambda_i),$$

where $\Lambda = \{\Lambda_1, \ldots, \Lambda_n\}$. In such a case, for a generic index i, we can assume q_j fixed for all $j \neq i$, and rewrite $\mathcal{L}(q, \Theta)$ as:

$$
\begin{aligned}
\mathcal{L}(q, \Theta) = & \int q_i(\mathbf{Z}_i|\Lambda_i) \left\{ \int \prod_{j \neq i} q_j(\mathbf{Z}_j|\Lambda_j) \log P(\mathbf{Z}, \mathcal{X}|\Theta)\, \mathrm{d}\mathbf{Z}_j \right\} \mathrm{d}\mathbf{Z}_i \\
& - \int q_i(\mathbf{Z}_i|\Lambda_i) \left\{ \int \prod_{j \neq i} q_j(\mathbf{Z}_j|\Lambda_j) \left\{ \log q_i(\mathbf{Z}_i|\Lambda_i) + \sum_{j \neq i} \log q_j(\mathbf{Z}_j|\Lambda_j) \right\} \mathrm{d}\mathbf{Z}_j \right\} \mathrm{d}\mathbf{Z}_i \\
= & \int q_i(\mathbf{Z}_i|\Lambda_i) \left\{ \int \prod_{j \neq i} q_j(\mathbf{Z}_j|\Lambda_j) \log P(\mathbf{Z}, \mathcal{X}|\Theta)\, \mathrm{d}\mathbf{Z}_j \right\} \mathrm{d}\mathbf{Z}_i \\
& - \int q_i(\mathbf{Z}_i|\Lambda_i) \left\{ \log q_i(\mathbf{Z}_i|\Lambda_i) \int \prod_{j \neq i} q_j(\mathbf{Z}_j|\Lambda_j)\, \mathrm{d}\mathbf{Z}_j \right\} \mathrm{d}\mathbf{Z}_i \\
& - \left\{ \int \prod_{j \neq i} q_j(\mathbf{Z}_j|\Lambda_j) \sum_{j \neq i} \log q_j(\mathbf{Z}_j|\Lambda_j)\, \mathrm{d}\mathbf{Z}_j \right\} \int q_i(\mathbf{Z}_i|\Lambda_i)\, \mathrm{d}\mathbf{Z}_i \\
= & \int q_i(\mathbf{Z}_i|\Lambda_i) \left\{ \int \prod_{j \neq i} q_j(\mathbf{Z}_j|\Lambda_j) \log P(\mathbf{Z}, \mathcal{X}|\Theta)\, \mathrm{d}\mathbf{Z}_j \right\} \mathrm{d}\mathbf{Z}_i \\
& - \int q_i(\mathbf{Z}_i|\Lambda_i) \log q_i(\mathbf{Z}_i|\Lambda_i)\, \mathrm{d}\mathbf{Z}_i \\
& + const.
\end{aligned}
$$

Consider now the term

$$
\int \prod_{j \neq i} q_j(\mathbf{Z}_j|\Lambda_j) \log P(\mathbf{Z}, \mathcal{X}|\Theta)\, \mathrm{d}\mathbf{Z}_j.
$$

By introducing a density function $f_i(\mathbf{Z}_i, \mathcal{X}|\Theta)$ defined by the relationship

$$
f_i(\mathbf{Z}_i, \mathcal{X}|\Theta) \propto \exp \left\{ \int \prod_{j \neq i} q_j(\mathbf{Z}_j|\Lambda_j) \log P(\mathbf{Z}, \mathcal{X}|\Theta)\, \mathrm{d}\mathbf{Z}_j \right\},
$$

we can rewrite the above expression as:

$$
\begin{aligned}
\mathcal{L}(q, \Theta) = & \int q_i(\mathbf{Z}_i|\Lambda_i) \left\{ \int \prod_{j \neq i} q_j(\mathbf{Z}_j|\Lambda_j) \log P(\mathbf{Z}, \mathcal{X}|\Theta)\, \mathrm{d}\mathbf{Z}_j \right\} \mathrm{d}\mathbf{Z}_i \\
& - \int q_i(\mathbf{Z}_i|\Lambda_i) \log q_i(\mathbf{Z}_i|\Lambda_i)\, \mathrm{d}\mathbf{Z}_i + const \\
\propto & \int q_i(\mathbf{Z}_i|\Lambda_i) \log f_i(\mathbf{Z}_i, \mathcal{X}|\Theta)\, \mathrm{d}\mathbf{Z}_i - \int q_i(\mathbf{Z}_i|\Lambda_i) \log q_i(\mathbf{Z}_i|\Lambda_i)\, \mathrm{d}\mathbf{Z}_i \\
= & -\mathbb{KL}(q_i, f_i).
\end{aligned}
$$

That is to say, the component $\mathcal{L}(q, \Theta)$ can be interpreted in terms of the Kullback-Leibler divergence between the factor q_i and the density f_i defined above, provided that all the remaining q_j components with $j \neq i$ are fixed. The minimal value for such divergence can be obtained when $q_i = f_i$, i.e., by solving the equation

$$q_i(\mathbf{Z}_i|\Lambda_i) = \frac{1}{W_i} \exp \left\{ \int \prod_{j \neq i} q_j(\mathbf{Z}_j|\Lambda_j) \log P(\mathbf{Z}, \mathcal{X}|\Theta) \, d\mathbf{Z}_j \right\}. \tag{A.9}$$

W_i represents a normalization factor, and, in general, we can choose a functional form for q_i such that Equation A.9 admits a closed solution for the variational parameter Λ_i. Dependencies amongst the Λ_i parameters can be solved by iterating the estimation, until convergence has been reached.

A.3   GIBBS SAMPLING

An alternative approach to variational approximation is to approach the problem of estimating $P(\mathbf{Z}|\mathcal{X}, \Theta)$ in a different, stochastic way, by using sampling techniques. Suppose that, although $P(\mathbf{Z}|\mathcal{X}, \Theta)$ cannot be computed analytically, it is possible to sample specific values $\{\mathbf{Z}^{(l)}\}_{l=1,\dots,L}$. Recall that, by Jensen inequality,

$$\log P(\mathcal{X}|\Theta) \geq \int P(\mathbf{Z}|\mathcal{X}, \Theta) \log P(\mathcal{X}, \mathbf{Z}|\Theta) \, d\mathbf{Z}.$$

The latter inequality suggests that we generalize the EM approach by defining the complete-data expectation log-likelihood,

$$\mathcal{Q}(\Theta, \Theta') = \int P(\mathbf{Z}|\mathcal{X}, \Theta') \log P(\mathcal{X}, \mathbf{Z}|\Theta) \, d\mathbf{Z},$$

and then alternating the optimization by inferring $P(\mathbf{Z}|\mathcal{X}, \Theta')$ given Θ' (E step), and subsequently by optimizing $\mathcal{Q}(\Theta, \Theta')$ with respect to Θ given $P(\mathbf{Z}|\mathcal{X}, \Theta')$. Since we are not able to infer $P(\mathbf{Z}|\mathcal{X}, \Theta')$, but we instead can sample $\{\mathbf{Z}^{(l)}\}_{l=1,\dots,L}$, we can approximate the complete-data expectation log likelihood as:

$$\mathcal{Q}(\Theta, \Theta') \approx \frac{1}{L} \sum_{l=1}^{L} \log P(\mathcal{X}, \mathbf{Z}^{(l)}|\Theta). \tag{A.10}$$

The latter does not depend on the posterior anymore, and hence can be approximated directly. In practice, the stochastic sample $\mathbf{Z}^{(l)}$ is used in place of the posterior distribution, thus resolving the complete data expectation likelihood in a stochastic version where the latent factors are statically assigned. We can hence devise a general *stochastic expectation maximization* procedure, as follows.

- Initialize $\Theta^{(0)}$;

- Iterate until convergence, for increasing steps t:

  (E step) For $l = 1, \ldots, L$ sample $\mathbf{Z}^{(l)} \sim P(\mathbf{Z}|\mathcal{X}, \Theta^{(t)})$;

  (M Step) Compute $\Theta^{(t+1)} = \text{argmax}_\Theta \, 1/L \sum_l \log P(\mathcal{X}, \mathbf{Z}^{(l)}|\Theta)$.

Clearly, the core of the approach is the capability of sampling $\mathbf{Z}^{(l)}$ from the posterior distribution. Several methods can be employed in this settings [31, 138]. We shall concentrate on Monte Carlo methods, and in particular with *Gibbs sampling*, which is a special case of *Markov chain Montecarlo* (MCMC) approximation [12]. MCMC methods approximate the probability $P(\mathbf{x})$ corresponding to a high-dimensional variable $\mathbf{x} \triangleq x_1, \ldots, x_n$ by a random walk on the state space governed by a Markov chain. The underlying idea of the Gibbs sampling procedure is the following. Assume an initial state $x_1^{(0)}, \ldots, x_n^{(0)}$ of the Markov chain. Each step of the Gibbs sampling procedure involves replacing the value of one of the variables by a value drawn from the distribution of that variable conditioned on the values of the remaining variables. That is, assuming that the probability $P(x_i|x_1, \ldots, x_{i-1}, x_{i+1}, \ldots, x_n)$ is tractable, we can compute the next state of the sampling process by iteratively sampling

$$x_i^{(t)} \sim P(x_i|x_1^{(t)}, \ldots, x_{i-1}^{(t)}, x_{i+1}^{(t-1)}, \ldots, x_n^{(t-1)}).$$

This procedure is repeated by cycling through the variables in some particular order.

To our purposes, the *Gibbs sampler* can be applied to estimate the posterior $P(\mathbf{Z}|\mathcal{X}, \Theta)$. In this case, assuming that $\mathbf{Z}$ can be decomposed into $\mathbf{z}_1, \ldots, \mathbf{z}_M$ and that $P(\mathbf{z}_i|\mathbf{z}_1, \ldots, \mathbf{z}_{i-1}, \mathbf{z}_{i+1}, \ldots, \mathbf{z}_M, \mathcal{X}, \Theta)$ is tractable, the E step of the above procedure can be refined into the following steps.

(Initialization) Randomly initialize $\mathbf{z}_1, \ldots, \mathbf{z}_M$;

[(Burn-in)] for $t = 1, \ldots, T$, iteratively compute

$$\mathbf{z}_i^{(t)} \sim P(\mathbf{z}_i|\mathbf{z}_1^{(t+1)}, \ldots, \mathbf{z}_{i-1}^{(t+1)}, \mathbf{z}_{i+1}^{(t)}, \ldots, \mathbf{z}_M^{(t)}, \mathcal{X}, \Theta)$$

to stabilize the initialization;

(E Step) Repeat the sampling process for each $\mathbf{z}_i$.

Bibliography

[1] R.P. Adams, G.E. Dahl, and I. Murray. Incorporating side information in probabilistic matrix factorization with gaussian processes. In *Proceedings of the 26th Conference on Uncertainty in Artificial Intelligence*, pages 1–9, 2010. 50

[2] D. Agarwal and B.C. Chen. Regression-based latent factor models. In *Proceedings of the 15th ACM SIGKDD International Conference on Knowledge Discovery and Data Mining*, KDD '09, pages 19–28, 2009. DOI: 10.1145/1557019.1557029. 112, 113, 114

[3] D. Agarwal and B.C. Chen. flda: matrix factorization through latent dirichlet allocation. In *Proceedings of the 6th ACM International Conference on Web Search and Data Mining*, WSDM '10, pages 91–100, 2010. DOI: 10.1145/1718487.1718499. 112, 115

[4] D. Agarwal and S. Merugu. Predictive discrete latent factor models for large scale dyadic data. In *Proceedings of the 13th ACM SIGKDD International Conference on Knowledge Discovery and Data Mining*, KDD '07, pages 26–35, 2007. DOI: 10.1145/1281192.1281199. 44

[5] C.C. Aggarwal, J.L. Wolf, K.L. Wu, and P.S. Yu. Horting hatches an egg: a new graph-theoretic approach to collaborative filtering. In *Proceedings of the 5th ACM SIGKDD International Conference on Knowledge Discovery and Data Mining*, KDD '99, pages 201–212, 1999. DOI: 10.1145/312129.312230. 9

[6] A. Ahmed, M. Aly, J. Gonzalez, S. Narayanamurthy, and A. J. Smola. Scalable inference in latent variable models. In *Proceedings of the 8th ACM International Conference on Web Search and Data Mining*, WSDM '12, pages 123–132, 2012. DOI: 10.1145/2124295.2124312. 85

[7] L. M. Aiello, A. Barrat, R. Schifanella, C. Cattuto, B. Markines, and F. Menczer. Friendship prediction and homophily in social media. *ACM Transactions on the Web*, 6(2):9:1–9:33, 2012. DOI: 10.1145/2180861.2180866. 128

[8] E. M. Airoldi, D. M. Blei, S. E. Fienberg, and E. P. Xing. Mixed membership stochastic blockmodels. *Journal of Machine Learning Research*, 9:1981–2014, 2008. 135

[9] M. Al Hasan and M.J. Zaki. A survey of link prediction in social networks. In *Social Network Data Analytics*, pages 243–275. Springer, 2011. DOI: 10.1007/978-1-4419-8462-3_9. 131

[10] L. AlSumait, D. Barbará, and C. Domeniconi. On-line lda: Adaptive topic models for mining text streams with applications to topic detection and tracking. In *Proceedings of the 8th IEEE International Conference on Data Mining*, ICDM '08, pages 3–12, 2008. DOI: 10.1109/ICDM.2008.140. 150

[11] A. Anagnostopoulos, R. Kumar, and M. Mahdian. Influence and correlation in social networks. In *Proceedings of the 14th ACM SIGKDD International Conference on Knowledge Discovery and Data Mining*, KDD '08, pages 7–15, 2008. DOI: 10.1145/1401890.1401897. 138

[12] C. Andrieu, N. de Freitas, A. Doucet, and Michael I. Jordan. An introduction to mcmc for machine learning. *Machine Learning*, 50(1-2):5–43, 2003. DOI: 10.1023/A:1020281327116. 159

[13] A. Asuncion, M. Welling, P. Smyth, and Y.W. Teh. On smoothing and inference for topic models. In *Proceedings of the 25th Conference on Uncertainty in Artificial Intelligence*, UAI '09, pages 27–34, 2009. 85

[14] R. Baeza-Yates and B. Ribeiro-Neto. *Modern Information Retrieval*. Addison-Wesley Longman Publishing Co., Inc., Boston, MA, USA, 1999. 5, 11, 12

[15] A. Banerjee and S. Basu. Topic models over text streams: A study of batch and online unsupervised learning. In *Proceedings of the 7th SIAM Conference on Data Mining*, SDM '07, pages 431–436, 2007. 150

[16] N. Barbieri. Regularized gibbs sampling for user profiling with soft constraints. In *Proceeding of the International Conference on Advances in Social Networks Analysis and Mining*, ASONAM '11, pages 129–136, 2011. DOI: 10.1109/ASONAM.2011.92. 65

[17] N. Barbieri, G. Costa, G. Manco, and R. Ortale. Modeling item selection and relevance for accurate recommendations: a bayesian approach. In *Proceedings of the 5th ACM Conference on Recommender Systems*, RecSys '11, pages 21–28, 2011. DOI: 10.1145/2043932.2043941. 65, 100

[18] N. Barbieri, G. Costa, G. Manco, and E. Ritacco. Characterizing relationships through co-clustering - a probabilistic approach. In *Proceedings of the International Conference on Knowledge Discovery and Information Retrieval*, KDIR '11, pages 64–73, 2011. 43, 44

[19] N. Barbieri and G. Manco. An analysis of probabilistic methods for top-n recommendation in collaborative filtering. In *Proceedings of the European Conference on Machine learning and Knowledge Discovery in Databases*, ECML/PKDD '11, pages 172–187, 2011. DOI: 10.1007/978-3-642-23780-5_21. 100

[20] N. Barbieri, G. Manco, R. Ortale, and E. Ritacco. Balancing prediction and recommendation accuracy: Hierarchical latent factors for preference data. In *Proceedings of the 12th SIAM International Conference on Data Mining*, SDM '12, pages 1035–1046, 2012. DOI: 10.1137/1.9781611972825.89. 73, 100

[21] N. Barbieri, G. Manco, and E. Ritacco. A probabilistic hierarchical approach for pattern discovery in collaborative filtering data. In *Proceedings of the 11th SIAM International Conference on Data Mining*, SDM '11, pages 630–621, 2011. DOI: 10.1137/1.9781611972818.54. 36

[22] N. Barbieri, G. Manco, E. Ritacco, M. Carnuccio, and A. Bevacqua. Probabilistic topic models for sequence data. *Machine Learning*, pages 1–25, 2013. DOI: 10.1007/s10994-013-5391-2. 124

[23] R. Bell, Y. Koren, and C. Volinsky. Modeling relationships at multiple scales to improve accuracy of large recommender systems. In *Proceedings of the 13th ACM SIGKDD International Conference on Knowledge Discovery and Data Mining*, KDD '07, pages 95–104, 2007. DOI: 10.1145/1281192.1281206. 16

[24] R.M. Bell and Y. Koren. Improved neighborhood-based collaborative filtering. In *Proceedings of the KDD Cup and Workshop in conjunction with KDD*, 2007. DOI: 10.1016/j.patrec.2011.10.016. 22

[25] R.M. Bell and Y. Koren. Scalable collaborative filtering with jointly derived neighborhood interpolation weights. In *Proceedings of the 7th IEEE International Conference on Data Mining*, ICDM '07, pages 43–52, 2007. DOI: 10.1109/ICDM.2007.90. 16

[26] Thiago Belluf, Leopoldo Xavier, and Ricardo Giglio. Case study on the business value impact of personalized recommendations on a large online retailer. In *Proceedings of the 6th ACM Conference on Recommender Systems*, RecSys '12, pages 277–280, 2012. DOI: 10.1145/2365952.2366014. 149

[27] J. Bennett, S. Lanning, and Netflix. The netflix prize. In *Proceedings of the KDD Cup and Workshop in conjunction with KDD*, 2007. 1

[28] A.L. Berger, S.A. Della Pietra, and V.J. Della Pietra. A maximum entropy approach to natural language processing. *Computational Linguistic*, 22(1):39–71, 1996. 28

[29] M.W. Berry, S.T. Dumais, and G.W. O'Brien. Using linear algebra for intelligent information retrieval. *SIAM Review*, 37(4):573–595, 1995. DOI: 10.1137/1037127. 18

[30] D. Billsus and M.J. Pazzani. Learning collaborative information filters. In *Proceedings of the 15th International Conference on Machine Learning*, ICML '98, pages 46–54, 1998. 18

[31] C. M. Bishop. *Pattern Recognition and Machine Learning*. Springer-Verlag New York, Inc., 2006. 23, 32, 53, 60, 81, 89, 117, 121, 123, 155, 159

[32] D. M. Blei. Introduction to probabilistic topic models. *Communications of the ACM*, 2011. DOI: 10.1145/2133806.2133826. 45

[33] D.M. Blei and J.D. Mcauliffe. Supervised topic models. In *Proceedings of the annual conference on Advances in Neural Information Processing Systems*, NIPS '08, 2008. 112

[34] D.M. Blei, A.Y. Ng, and M.I. Jordan. Latent dirichlet allocation. *Journal of Machine Learning Research*, 3(1):993–1022, 2003. 45, 46, 61, 62

[35] F. Bonchi. Influence propagation in social networks: A data mining perspective. *IEEE Intelligent Informatics Bulletin*, 12(1):8–16, 2011. DOI: 10.1109/WI-IAT.2011.286. 137

[36] J.S. Breese, D. Heckerman, and C. Kadie. Empirical analysis of predictive algorithms for collaborative filtering. In *Proceedings of the 14th Conference on Uncertainty in Artificial Intelligence*, UAI '98, pages 43–52, 1998. 14

[37] I. Cadez, D. Heckerman, C. Meek, P. Smyth, and S. White. Visualization of navigation patterns on a web site using model-based clustering. In *Proceedings of the 6th ACM SIGKDD International Conference on Knowledge Discovery and Data Mining*, KDD '00, pages 280–284, 2000. 118

[38] J. Canny. Collaborative filtering with privacy. In *Proceedings of the IEEE Symposium on Security and Privacy*, SP '02, pages 45+, 2002. DOI: 10.1145/347090.347151. 9, 10

[39] J. Canny. Collaborative filtering with privacy via factor analysis. In *Proceedings of the 25th ACM SIGIR Conference on Research and Development in Information Retrieval*, SIGIR '02, pages 238–245, 2002. DOI: 10.1145/564376.564419. 10

[40] M. Caraciolo. Recommendations and how to measure the roi with some metrics ? 149

[41] M. Cha, H. Haddadi, F. Benevenuto, and K.P. Gummadi. Measuring user influence in twitter: The million follower fallacy. In *Proceedings of the 4th AAAI Conference on Weblogs and Social Media*, ICWSM '10, 2010. 140

[42] Youngchul Cha and Junghoo Cho. Social-network analysis using topic models. In *Proceedings of the 35th ACM SIGIR Conference on Research and Development in Information Retrieval*, SIGIR '12, pages 565–574, 2012. DOI: 10.1145/2348283.2348360. 134

[43] W. Chen, C. Wang, and Y. Wang. Scalable influence maximization for prevalent viral marketing in large-scale social networks. In *Proceedings of the 16th ACM SIGKDD International Conference on Knowledge Discovery and Data Mining*, KDD '10, pages 1029–1038, 2010. DOI: 10.1145/1835804.1835934. 140

[44] Y. Chen, W. Peng, and S. Lee. Efficient algorithms for influence maximization in social networks. *Knowledge and Information Systems*, 33(3):577–601, 2012. DOI: 10.1007/s10115-012-0540-7. 140

[45] T.M. Cover and J.A. Thomas. *Elements of Information Theory*. Wiley-Interscience, New York, NY, USA, 2006. 30, 153, 156

[46] D. Crandall, D. Cosley, D. Huttenlocher, J. Kleinberg, and S. Suri. Feedback effects between similarity and social influence in online communities. In *Proceedings of the 14th ACM SIGKDD International Conference on Knowledge Discovery and Data Mining*, KDD '08, pages 160–168, 2008. DOI: 10.1145/1401890.1401914. 128

[47] P. Cremonesi, Y. Koren, and R. Turrin. Performance of recommender algorithms on top-n recommendation tasks. In *Proceedings of the 4th ACM Conference on Recommender Systems*, RecSys '10, pages 39–46, 2010. DOI: 10.1145/1864708.1864721. 6, 87, 97, 98

[48] J. N. Darroch and D. Ratcliff. Generalized iterative scaling for log-linear models. *The Annals of Mathematical Statistics*, 43:1470–1480, 1972. DOI: 10.1214/aoms/1177692379. 29

[49] A.S. Das, M. Datar, A. Garg, and S. Rajaram. Google news personalization: scalable online collaborative filtering. In *Proceedings of the 16th international conference on World Wide Web*, WWW '07, pages 271–280, 2007. DOI: 10.1145/1242572.1242610. 1

[50] S.C Deerwester, S.T. Dumais, T.K. Landauer, G.W. Furnas, and R.A. Harshman. Indexing by latent semantic analysis. *Journal of the American Society for Information Science*, 41(6):391–407, 1990. DOI: 10.1002/(SICI)1097-4571(199009)41:6%3C391::AID-ASI1%3E3.0.CO;2-9. 17

[51] A.P. Dempster, N.M. Laird, and D.B. Rubin. Maximum likelihood from incomplete data via the em algorithm. *Journal of the Royal Statistical Society. Series B*, 39(1):1–38, 1977. 34

[52] I.S. Dhillon and D.S. Modha. Concept decompositions for large sparse text data using clustering. *Machine Learning*, 42(1-2):143–175, 2001. DOI: 10.1023/A:1007612920971. 12

[53] O. Domingos and M. Richardson. Mining the network value of customers. In *Proceedings of the 7th ACM SIGKDD International Conference on Knowledge Discovery and Data Mining*, KDD '01, pages 57–66, 2001. DOI: 10.1145/502512.502525. 139

[54] L. Du, W. L. Buntine, and H. Jin. Sequential latent dirichlet allocation: Discover underlying topic structures within a document. In *Proceedings of the 10th IEEE International Conference on Data Mining*, ICDM '10, pages 148–157, 2010. DOI: 10.1109/ICDM.2010.51. 123

[55] E. Erosheva, S. Fienberg, and J. Lafferty. Mixed-membership models of scientific publications. *Proceedings of the National Academy of Science*, 101:5220–5227, 2004. DOI: 10.1073/pnas.0307760101. 132

[56] T. Fawcett. An introduction to roc analysis. *Pattern Recognition Letters*, 27(8):861–874, 2006. DOI: 10.1016/j.patrec.2005.10.010. 5

[57] M.A.T. Figueiredo and A.K. Jain. Unsupervised learning of finite mixture models. *IEEE Transactions on Pattern Analysis and Machine Intelligence*, 24(3):381–396, 2002. DOI: 10.1109/34.990138. 57

[58] D. Fink. A compendium of conjugate priors, 1997. 54

[59] N. E. Friedkin. *A Structural Theory of Social Influence*. Cambridge University Press, 1998. DOI: 10.1017/CBO9780511527524. 128

[60] S. Funk. Netflix update: Try this at home. URL: http://sifter.org/ simon/Journal/20061211.html, 2006. 19, 20

[61] M. Ge, C. Delgado-Battenfeld, and D. Jannach. Beyond accuracy: evaluating recommender systems by coverage and serendipity. In *Proceedings of the 4th ACM Conference on Recommender Systems*, RecSys '10, pages 257–260, 2010. DOI: 10.1145/1864708.1864761. 149

[62] A. Gelman, J.B. Carlin, H.S. Stern, and D.B. Rubin. *Bayesian Data Analysis*. Chapman & ofHall/CRC Press, 2004. 53, 62

[63] T. George and S. Merugu. A scalable collaborative filtering framework based on coclustering. In *Proceedings of the 5th IEEE International Conference on Data Mining*, ICDM '05, pages 625–628, 2005. DOI: 10.1109/ICDM.2005.14. 40

[64] S.J. Gershman and D.M. Blei. A tutorial on bayesian nonparametric models. *Journal of Mathematical Psychology*, 56(1):1–12, 2012. DOI: 10.1016/j.jmp.2011.08.004. 150

[65] M. Girolami and A. Kabán. On an equivalence between plsi and lda. In *Proceedings of the 26th ACM SIGIR Conference on Research and Development in Information Retrieval*, SIGIR '03, pages 433–434, 2003. DOI: 10.1145/860435.860537. 46

[66] Malcolm Gladwell. *The tipping point: how little things can make a big difference*. Little Brown, 2000. 137

[67] A. Gohr, A. Hinneburg, R. Schult, and M. Spiliopoulou. Topic evolution in a stream of documents. In *Proceedings of the 9th SIAM Conference on Data Mining*, SDM '09, pages 859–872, 2009. 150

[68] P. Gopalan, D.M. Mimno, S. Gerrish, M.J. Freedman, and D. M. Blei. Scalable inference of overlapping communities. In *Proceedings of the annual conference on Advances in Neural Information Processing Systems*, NIPS '12, pages 2258–2266, 2012. 137

[69] G. Govaert and M. Nadif. Clustering with block mixture models. *Pattern Recognition*, 36(2):463–473, 2003. DOI: 10.1016/S0031-3203(02)00074-2. 42, 43

[70] G. Govaert and M. Nadif. An em algorithm for the block mixture model. *IEEE Transactions on Pattern Analysis and Machine Intelligence*, 27(4):643–647, 2005. DOI: 10.1109/TPAMI.2005.69. 40, 42

[71] A. Goyal, F. Bonchi, and L. Lakshmanan. Learning influence probabilities in social networks. In *Proceedings of the 6th ACM International Conference on Web Search and Data Mining*, WSDM '10, pages 241–250, 2010. DOI: 10.1145/1718487.1718518. 141

[72] A. Goyal, F. Bonchi, and L.V.S. Lakshmanan. A data-based approach to social influence maximization. *Proceedings of the VLDB Endowment*, 5(1):73–84, 2011. 140

[73] A. Goyal and L. Lakshmanan. Recmax: exploiting recommender systems for fun and profit. In *Proceedings of the 18th ACM SIGKDD International Conference on Knowledge Discovery and Data Mining*, KDD '12, pages 1294–1302, 2012. DOI: 10.1145/2339530.2339731. 144

[74] T. L. Griffiths, M. Steyvers, and J. B. Tenenbaum. Topics in semantic representation. *Psychological Review*, 114, 2007. DOI: 10.1037/0033-295X.114.2.211. 123

[75] T.L. Griffiths and M. Steyvers. Finding scientific topics. *Proceedings of the National Academy of Sciences*, 101:5228–5235, 2004. DOI: 10.1073/pnas.0307752101. 62

[76] A. Gruber, Y. Weiss, and M. Rosen-Zvi. Hidden topic markov models. *Journal of Machine Learning Research*, 2:162–170, 2007. 123

[77] G. Heinrich. Parameter estimation for text analysis. Technical report, University of Leipzig, 2008. 62

[78] J. Herlocker, J.A. Konstan, and J. Riedl. An empirical analysis of design choices in neighborhood-based collaborative filtering algorithms. *Information Retrieval*, 5(4):287–310, 2002. DOI: 10.1023/A:1020443909834. 15

[79] J.L. Herlocker, J. A. Konstan, L.G. Terveen, and J. T. Riedl. Evaluating collaborative filtering recommender systems. *ACM Transactions on Information Systems*, 22(1):5–53, 2004. DOI: 10.1145/963770.963772. 4

[80] M.D. Hoffman, D.M. Blei, and F.R. Bach. Online learning for latent dirichlet allocation. In *Proceedings of the annual conference on Advances in Neural Information Processing Systems*, NIPS '10, 2010. 85

[81] T. Hofmann. Probabilistic latent semantic indexing. In *Proceedings of the 22nd ACM SIGIR Conference on Research and Development in Information Retrieval*, SIGIR '99, pages 50–57, 1999. DOI: 10.1145/312624.312649. 46, 47

[82] T. Hofmann. Learning what people (don't) want. In *Proceedings of the 12th European Conference on Machine Learning*, ECML '01, pages 214–225. Springer-Verlag, 2001. DOI: 10.1023/A:1007617005950. 48

[83] T. Hofmann. Unsupervised learning by probabilistic latent semantic analysis. *Machine Learning*, 42(1):177–196, 2001. DOI: 10.1023/A:1007617005950. 46

[84] T. Hofmann. Collaborative filtering via gaussian probabilistic latent semantic analysis. In *Proceedings of the 26th ACM SIGIR Conference on Research and Development in Information Retrieval*, SIGIR '03, pages 259–266, 2003. DOI: 10.1145/860435.860483. 48

[85] T. Hofmann. Latent semantic models for collaborative filtering. *ACM Transactions on Information Systems*, 22(1):89–115, 2004. DOI: 10.1145/963770.963774. 27

[86] T. Hofmann and J. Puzicha. Latent class models for collaborative filtering. In *Proceedings of the 16th International Joint Conference on Artificial Intelligence*, IJCAI '99, pages 688–693. Morgan Kaufmann Publishers Inc., 1999. 9, 38, 41

[87] L. Hong, A. Ahmed, S. Gurumurthy, A.J. Smola, and K. Tsioutsiouliklis. Discovering geographical topics in the twitter stream. In *Proceedings of the 21st International Conference on World Wide Web*, WWW '12, pages 769–778, 2012. DOI: 10.1145/2187836.2187940. 149

[88] P.O. Hoyer. Non-negative matrix factorization with sparseness constraints. *J. Mach. Learn. Res.*, 5:1457–1469, 2004. 20

[89] F. Huang, C. Hsieh, K. Chan, and C. Lin. Iterative scaling and coordinate descent methods for maximum entropy models. *Journal of Machine Learning Research*, 11:815–848, 2010. 31

[90] N.J. Hurley, M.P. O'Mahony, and G.C.M. Silvestre. Attacking recommender systems: A cost-benefit analysis. *IEEE Intelligent Systems*, 22(3):64–68, 2007. DOI: 10.1109/MIS.2007.44. 150

[91] A. Ilin and T. Raiko. Practical approaches to principal component analysis in the presence of missing values. *Journal of Machine Learning Research*, 11(1):1957–2000, 2010. 80

[92] T. Iwata, T. Yamada, Y. Sakurai, and N Ueda. Online multiscale dynamic topic models. In *Proceedings of the 16th ACM SIGKDD International Conference on Knowledge Discovery and Data Mining*, KDD '10, pages 663–672, 2010. DOI: 10.1145/1835804.1835889. 150

[93] A.K Jain, M.N. Murty, and P.J. Flynn. Data clustering: a review. *ACM Computing Surveys*, 31(3):264–323, 1999. DOI: 10.1145/331499.331504. 12

[94] M. Jamali and M. Ester. Trustwalker: a random walk model for combining trust-based and item-based recommendation. In *Proceedings of the 15th ACM SIGKDD International Conference on Knowledge Discovery and Data Mining*, KDD '09, pages 397–406, 2009. DOI: 10.1145/1557019.1557067. 131

[95] M. Jamali, G. Haffari, and M. Ester. Modeling the temporal dynamics of social rating networks using bidirectional effects of social relations and rating patterns. In *Proceedings of the 20th International Conference on World Wide Web*, WWW '11, pages 527–536, 2011. DOI: 10.1145/1963405.1963480. 130, 131

[96] M. Jamali, T. Huang, and M. Ester. A generalized stochastic block model for recommendation in social rating networks. In *Proceedings of the 5th ACM Conference on Recommender Systems*, RecSys '11, pages 53–60, 2011. DOI: 10.1145/2043932.2043946. 135

[97] E. T. Jaynes. Prior probabilities. *IEEE Transactions on Systems Science and Cybernetics*, 4(3):227–241, 1968. DOI: 10.1109/TSSC.1968.300117. 57

[98] R. Jin, L. Si, and C. Zhai. A study of mixture models for collaborative filtering. *Information Retrieval*, 9(3):357–382, 2006. DOI: 10.1007/s10791-006-4651-1. 41

[99] R. Jin, R. Yan, Z. Jian, and A.G. Hauptmann. A faster iterative scaling algorithm for conditional exponential model. In *Proceedings of the 20th International Conference on Machine Learning*, ICML'03, pages 282–289, 2003. 31

[100] X. Jin, Y. Zhou, and B. Mobasher. Web usage mining based on probabilistic latent semantic analysis. In *Proceedings of the 10th ACM SIGKDD International Conference on Knowledge Discovery and Data Mining*, KDD '04, pages 197–205, 2004. DOI: 10.1145/1014052.1014076. 101, 102

[101] M.I. Jordan. Graphical models. *Statistical Science*, 19:140–155, 2004. DOI: 10.1214/088342304000000026. 25

[102] M.I. Jordan, Z. Ghahramani, T.S. Jaakkola, and L.K. Saul. An introduction to variational methods for graphical models. *Machine learning*, 37(2):183–233, 1999. DOI: 10.1023/A:1007665907178. 155

[103] G. Karypis. Evaluation of item-based top-n recommendation algorithms. In *Proceedings of the 10th International Conference on Information and Knowledge Management*, CIKM '01, pages 247–254, 2001. DOI: 10.1145/502585.502627. 4

[104] D. Kempe, J. Kleinberg, and É. Tardos. Maximizing the spread of influence through a social network. In *Proceedings of the 9th ACM SIGKDD International Conference on Knowledge Discovery and Data Mining*, KDD '03, pages 137–146, 2003. DOI: 10.1145/956750.956769. 139, 140

[105] M. Kimura and K. Saito. Tractable models for information diffusion in social networks. In *Proceedings of the European Conference on Machine learning and Knowledge Discovery in Databases*, volume 4213 of *ECML/PKDD '06*, pages 259–271, 2006. DOI: 10.1007/11871637_27. 140

[106] Y. Koren. Factorization meets the neighborhood: a multifaceted collaborative filtering model. In *Proceedings of the 14th ACM SIGKDD International Conference on Knowledge Discovery and Data Mining*, KDD '08, pages 426–434, 2008. DOI: 10.1145/1401890.1401944. 21

[107] Y. Koren, R. Bell, and C. Volinsky. Matrix factorization techniques for recommender systems. *IEEE Computer*, 42(8):30–37, 2009. DOI: 10.1109/MC.2009.263. 21

[108] B. Krulwich and C. Burkey. Lifestyle finder: Intelligent user profiling using large-scale demographic data. *AI Magazine*, 18(2):37–45, 1997. DOI: 10.1609/aimag.v18i2.1292. 11

[109] S.K. Lam and J. Riedl. Shilling recommender systems for fun and profit. In *Proceedings of the 13th International Conference on World Wide Web*, WWW '04, pages 393–402, 2004. DOI: 10.1145/988672.988726. 10

[110] K. Lang. Newsweeder: Learning to filter netnews. In *Proceedings of the 17th International Conference on Machine Learning*, ICML '00, 2000. 11

[111] N. Lathia, S. Hailes, L. Capra, and X. Amatriain. Temporal diversity in recommender systems. In *Proceedings of the 33rd ACM SIGIR Conference on Research and Development in Information Retrieval*, SIGIR '10, pages 210–217, 2010. DOI: 10.1145/1835449.1835486. 149

[112] D.D. Lee and H.S. Seung. Learning the parts of objects by nonnegative matrix factorization. *Nature*, 401:788–791, 1999. DOI: 10.1038/46985. 20

[113] J. Leskovec, A. Krause, C. Guestrin, C. Faloutsos, J. VanBriesen, and N. Glance. Cost-effective outbreak detection in networks. In *Proceedings of the 13th ACM SIGKDD International Conference on Knowledge Discovery and Data Mining*, KDD '07, pages 420–429, 2007. DOI: 10.1145/1281192.1281239. 140

[114] H. Lieberman. Letizia: An agent that assists web browsing. In *Proceedings of the 12th International Joint Conference on Artificial Intelligence*, IJCAI '95, pages 924 – 929, 1995. 13

[115] Y.W. Lim and Y.W. Teh. Variational bayesian approach to movie rating prediction. In *Proceedings of KDD Cup and Workshop in conjunction with KDD*, 2007. 80

[116] D. C. Liu and J. Nocedal. On the limited memory bfgs method for large scale optimization. *Mathematical Programming*, 45(3):503–528, 1989. DOI: 10.1007/BF01589116. 31

[117] L. Liu, J. Tang, J. Han, M. Jiang, and S. Yang. Mining topic-level influence in heterogeneous networks. In *Proceedings of the 19th ACM International Conference on Information and Knowledge Management*, CIKM '10, pages 199–208, 2010. DOI: 10.1145/1871437.1871467. 143

[118] H. Ma, H. Yang, M.R. Lyu, and I. King. Sorec: social recommendation using probabilistic matrix factorization. In *Proceedings of the 17th ACM International Conference on Information and Knowledge Management*, CIKM '08, pages 931–940, 2008. DOI: 10.1145/1458082.1458205. 134

[119] H. Ma, D. Zhou, C. Liu, M. R. Lyu, and I. King. Recommender systems with social regularization. In *Proceedings of the 7th ACM International Conference on Web Search and Data Mining*, WSDM '11, pages 287–296, 2011. DOI: 10.1145/1935826.1935877. 132

[120] R. Malouf. A comparison of algorithms for maximum entropy parameter estimation. In *Proceedings of the 6th Conference on Natural Language Learning, CoNLL*, CoNLL '02, pages 1–7, 2002. DOI: 10.3115/1118853.1118871. 31

[121] U. Manber, A. Patel, and J. Robison. Experience with personalization on yahoo! *Communications of the ACM*, 43(8):35–39, 2000. DOI: 10.1145/345124.345136. 11

[122] C. D. Manning and H. Schütze. *Foundations of Statistical Natural Language Processing*. MIT Press, 1999. 121

[123] B. Marlin. Modeling user rating profiles for collaborative filtering. In *Proceedings of the annual conference on Advances in Neural Information Processing Systems*, NIPS '03, 2003. 65

[124] B. Marlin. Collaborative filtering: A machine learning perspective. Technical report, Department of Computer Science University of Toronto, 2004. 36

[125] P. Massa and P. Avesani. Trust-aware bootstrapping of recommender systems. In *Proceedings of the ECAI Workshop on Recommender Systems*, pages 29–33, 2006. 129

[126] S. M. McNee, J. Riedl, and J.A. Konstan. Being accurate is not enough: how accuracy metrics have hurt recommender systems. In *Proceedings of the CHI '06 Extended Abstracts on Human Factors in Computing Systems*, CHI EA '06, pages 1097–1101, 2006. DOI: 10.1145/1125451.1125659. 87, 108, 149

172 BIBLIOGRAPHY

[127] M. McPherson, L. Smith-Lovin, and J. M. Cook. Birds of a feather: Homophily in social networks. *Annual Review of Sociology*, 27(1):415–444, 2001. DOI: 10.1146/annurev.soc.27.1.415. 128

[128] Q. Mei, D. Cai, D. Zhang, and C. Zhai. Topic modeling with network regularization. In *Proceedings of the 17th international conference on World Wide Web*, WWW '08, pages 101–110, 2008. DOI: 10.1145/1367497.1367512. 132

[129] Q. Mei, X. Shen, and C. Zhai. Automatic labeling of multinomial topic models. In *Proceedings of the 13th ACM SIGKDD International Conference on Knowledge Discovery and Data Mining*, KDD '07, pages 490–499, 2007. DOI: 10.1145/1281192.1281246. 102

[130] A. K. Menon, X. Jiang, S. Vembu, C. Elkan, and L. Ohno-Machado. Predicting accurate probabilities with a ranking loss. In *Proceedings of the 29th International Conference on Machine Learning*, ICML '12, 2012. 89

[131] A.K. Menon and C. Elkan. A log-linear model with latent features for dyadic prediction. In *Proceedings of the 10th IEEE International Conference on Data Mining*, ICDM '10, pages 364–373, 2010. DOI: 10.1109/ICDM.2010.148. 28, 89, 112

[132] A.K. Menon and C. Elkan. Fast algorithms for approximating the singular value decomposition. *ACM Transactions on Knowledge Discovery in Databases*, 5(2):1–1:36, 2011. DOI: 10.1145/1921632.1921639. 18

[133] A.K. Menon and C. Elkan. Link prediction via matrix factorization. In *Proceedings of the European Conference on Machine learning and Knowledge Discovery in Databases*, ECML/PKDD '11, pages 437–452, 2011. DOI: 10.1007/978-3-642-23783-6_28. 21

[134] T. Minka and J. Lafferty. Expectation-propagation for the generative aspect model. In *Proceedings of the 18th Conference on Uncertainty in Artificial Intelligence*, UAI '02, pages 352–359, 2002. 62

[135] T.P. Minka. Estimating a dirichlet distribution. Technical report, Microsoft Research, 2000. 64

[136] B. Mobasher, R. Burke, R. Bhaumik, and C. Williams. Toward trustworthy recommender systems: An analysis of attack models and algorithm robustness. *ACM Transactions on Internet Technology*, 7(4):23, 2007. DOI: 10.1145/1278366.1278372. 10, 150

[137] T. Murakami, K. Mori, and R. Orihara. Metrics for evaluating the serendipity of recommendation lists. In *Proceedings of the Conference on New frontiers in artificial intelligence*, JSAI '07, pages 40–46, 2008. DOI: 10.1007/978-3-540-78197-4_5. 149

[138] K.P. Murphy. *Machine Learning: A Probabilistic Perspective*. The MIT Press, Cambridge, MA, USA, 2012. 23, 88, 155, 159

[139] G. Manco N. Barbieri, F. Bonchi. Topic-aware social influence propagation models. In *Proceedings of the 12nd IEEE International Conference on Data Mining*, ICDM '12, pages 81–90, 2012. DOI: 10.1109/ICDM.2012.122. 143

[140] R. Nallapati, W. Cohen, and J. Lafferty. Parallelized variational em for latent dirichlet allocation: An experimental evaluation of speed and scalability. In *Proceedings of the Seventh IEEE ICDM Conference Workshops*, pages 349–354, 2007. DOI: 10.1109/ICDMW.2007.70. 84

[141] A. Narayanan and V. Shmatikov. How to break anonymity of the netflix prize dataset. *CoRR*, abs/cs/0610105, 2006. 10

[142] G. L. Nemhauser, L. A. Wolsey, and M. L. Fisher. An analysis of approximations for maximizing submodular set functions - i. *Mathematical Programming*, 14(1):265–294, 1978. DOI: 10.1007/BF01588971. 140

[143] D. Newman, A. Asuncion, P. Smyth, and M. Welling. Distributed algorithms for topic models. *Journal of Machine Learning Research*, 10:1801–1828, 2009. 85

[144] J. Nocedal and S.J. Wright. *Numerical optimization*. Springer series in operations research and financial engineering. Springer, 2006. 31

[145] M. O'Mahony, N. Hurley, N. Kushmerick, and G. Silvestre. Collaborative recommendation: A robustness analysis. *ACM Transactions on Internet Technology*, 4(4):344–377, 2004. DOI: 10.1145/1031114.1031116. 10

[146] M. Papagelis, D. Plexousakis, and T. Kutsuras. Alleviating the sparsity problem of collaborative filtering using trust inferences. In P. Herrmann, V. Issarny, and S. Shiu, editors, *New Trends in Applied Artificial Intelligence*, volume 3477 of *Lecture Notes in Computer Science*, pages 224–239. Springer Berlin Heidelberg, 2005. 9

[147] E. Pariser. *The filter bubble : what the Internet is hiding from you*. Penguin Press, 2011. 108

[148] A. Paterek. Improving regularized singular value decomposition for collaborative filtering. In *Proceedings of the KDD Cup and Workshop in conjunction with KDD*, 2007. 21

[149] D. Pavlov, E. Manavoglu, D.M. Pennock, and C.L. Giles. Collaborative filtering with maximum entropy. *IEEE Intelligent Systems*, 19(6):40–48, 2004. DOI: 10.1109/MIS.2004.59. 28

[150] M. Pazzani, D. Billsus, S. Michalski, and J. Wnek. Learning and revising user profiles: The identification of interesting web sites. *Machine Learning*, 27(3):313–331, 1997. DOI: 10.1023/A:1007369909943. 11

[151] M.J. Pazzani. A framework for collaborative, content-based and demographic filtering. *Artificial Intelligence Review*, 13(5):393–408, 1999. DOI: 10.1023/A:1006544522159. 9

[152] M.J. Pazzani and D. Billsus. Content-based recommendation systems. In P. Brusilovsky, A. Kobsa, and W. Nejdl, editors, *The Adaptive Web*, pages 325–341. Springer-Verlag, Berlin, Heidelberg, 2007. DOI: 10.1007/978-3-540-72079-9. 13

[153] H. Polat and W. Du. Privacy-preserving collaborative filtering using randomized perturbation techniques. In *Proceedings of the 3rd IEEE International Conference on Data Mining*, ICDM '03, pages 625–628, 2003. DOI: 10.1109/ICDM.2003.1250993. 10

[154] A. Popescul, L. Ungar, D. Pennock, and S. Lawrence. Probabilistic models for unified collaborative and content-based recommendation in sparse-data environments. In *Proceedings of the 17th Conference on Uncertainty in Artificial Intelligence*, UAI' 01, pages 437–444, 2001. 9

[155] I. Porteous, E. Bart, and M. Welling. Multi-hdp: a non parametric bayesian model for tensor factorization. In *Proceedings of the National Conference of the American Association for Artificial Intelligence*, AAAI '08, pages 1487–1490, 2008. 70

[156] I. Porteous, D. Newman, A. Ihler, A. Asuncion, P. Smyth, and M. Welling. Fast collapsed gibbs sampling for latent dirichlet allocation. In *Proceedings of the 14th ACM SIGKDD International Conference on Knowledge Discovery and Data Mining*, KDD '08, pages 569–577, 2008. DOI: 10.1145/1401890.1401960. 84, 85

[157] N. Ramakrishnan, B.J. Keller, B.J. Mirza, A.Y. Grama, and G. Karypis. Privacy risks in recommender systems. *IEEE Internet Computing*, 5(6):54–62, 2001. DOI: 10.1109/4236.968832. 10

[158] A.M. Rashid, I. Albert, D. Cosley, S.K. Lam, S.M. McNee, J.A. Konstan, and J. Riedl. Getting to know you: learning new user preferences in recommender systems. In *Proceedings of the 7th International Conference on Intelligent User Interfaces*, IUI '10, pages 127–134, 2002. DOI: 10.1145/502716.502737. 9

[159] S. Rendle, C. Freudenthaler, Z. Gantner, and L. Schmidt-Thieme. Bpr: Bayesian personalized ranking from implicit feedback. In *Proceedings of the 25th Conference on Uncertainty in Artificial Intelligence*, UAI '09, pages 452–461, 2009. 89

[160] J.D. M. Rennie and N. Srebro. Fast maximum margin matrix factorization for collaborative prediction. In *Proceedings of the 22nd International Conference on Machine Learning*, ICML '05, pages 713–719, 2005. DOI: 10.1145/1102351.1102441. 20

[161] P. Resnick, N. Iacovou, M. Suchak, P. Bergstrom, and J. Riedl. Grouplens: An open architecture for collaborative filtering of netnews. In *Proceedings of the ACM Conference on*

Computer Supported Cooperative Work, CSCW '94, pages 175–186, New York, NY, USA, 1994. ACM. DOI: 10.1145/192844.192905. 69

[162] F. Ricci, L. Rokach, B. Shapira, and P.B. Kantor, editors. *Recommender Systems Handbook.* Springer-Verlag, New York, NY, USA, 2010. 1

[163] M. Richardson and P. Domingos. Mining knowledge-sharing sites for viral marketing. In *Proceedings of the 8th ACM SIGKDD International Conference on Knowledge Discovery and Data Mining*, KDD '02, pages 61–70, 2002. DOI: 10.1145/775047.775057. 139

[164] A. Saha and V. Sindhwani. Learning evolving and emerging topics in social media: a dynamic nmf approach with temporal regularization. In *Proceedings of the 8th ACM International Conference on Web Search and Data Mining*, WSDM '12, pages 693–702, 2012. DOI: 10.1145/2124295.2124376. 150

[165] K. Saito, M. Kimura, K. Ohara, and H. Motoda. Efficient discovery of influential nodes for sis models in social networks. *Knowledge and Information Systems*, 30(3):613–635, 2012. DOI: 10.1007/s10115-011-0396-2. 140

[166] K. Saito, R. Nakano, and M. Kimura. Prediction of information diffusion probabilities for independent cascade model. In *Proceedings of the 12th International Conference on Knowledge-Based Intelligent Information and Engineering Systems*, KES '08, pages 67–75, 2008. DOI: 10.1007/978-3-540-85567-5_9. 141

[167] Patty Sakunkoo and Nathan Sakunkoo. Analysis of social influence in online book reviews. In *Proceedings of the 3rd International Conference on Weblogs and Social Media*, ICSWM '09, 2009. 129

[168] R. Salakhutdinov and A. Mnih. Bayesian probabilistic matrix factorization using markov chain monte carlo. In *Proceedings of the 25th International Conference on Machine Learning*, ICML '08, pages 880–887, 2008. DOI: 10.1145/1390156.1390267. 50, 81, 125

[169] R. Salakhutdinov and A. Mnih. Probabilistic matrix factorization. In *Proceedings of the annual conference on Advances in Neural Information Processing Systems*, NIPS '08, pages 1257–1264, 2008. 49, 50

[170] G. Salton and M.J. McGill. *Introduction to Modern Information Retrieval.* McGraw-Hill, Inc., New York, NY, USA, 1986. 11, 12

[171] J.J. Sandvig, B. Mobasher, and R. Burke. Robustness of collaborative recommendation based on association rule mining. In *Proceedings of the 1st ACM Conference on Recommender Systems*, RecSys '07, pages 105–112, 2007. DOI: 10.1145/1297231.1297249. 10

[172] R. L.T. Santos, C. Macdonald, and I. Ounis. Exploiting query reformulations for web search result diversification. In *Proceedings of the 19th International Conference on World Wide Web*, WWW '10, pages 881–890, 2010. DOI: 10.1145/1772690.1772780. 109

[173] B. Sarwar, G. Karypis, J. Konstan, and J. Riedl. Incremental singular value decomposition algorithms for highly scalable recommender systems. In *Proceedings of the 5th International Conference on Computer and Information Science*, ICIS '02, pages 27–28, 2002. 9

[174] B.M. Sarwar, G. Karypis, J.A. Konstan, and J. Riedl. Item-based collaborative filtering recommendation algorithms. In *Proceedings of the 10th International Conference on World Wide Web*, WWW '01, pages 285–295, 2001. DOI: 10.1145/371920.372071. 5, 9, 15, 87

[175] B.M. Sarwar, G. Karypis, J.A. Konstan, and J.T. Riedl. Application of dimensionality reduction in recommender systems: A case study. In *Proceedings of the ACM WebKDD Workshop*, 2000. 9, 18

[176] J.B. Schafer, J.A. Konstan, and J. Riedl. E-commerce recommendation applications. *Data Mining and Knowledge Discovery*, 5(1-2):115–153, 2001. DOI: 10.1023/A:1009804230409. 1

[177] A.I. Schein, A. Popescul, L.H. Ungar, and D.M. Pennock. Methods and metrics for cold-start recommendations. In *Proceedings of the 25th ACM SIGIR Conference on Research and Development in Information Retrieval*, SIGIR '02, pages 253–260, 2002. DOI: 10.1145/564376.564421. 9

[178] M.M. Shafiei and E.E. Milios. Latent dirichlet co-clustering. In *Proceedings of the 6th IEEE International Conference on Data Mining*, ICDM '06, pages 542–551, 2006. DOI: 10.1109/ICDM.2006.94. 70

[179] H. Shan and A. Banerjee. Bayesian co-clustering. In *Proceedings of the 8th IEEE International Conference on Data Mining*, ICDM '08, pages 530–539, 2008. DOI: 10.1109/ICDM.2008.91. 70

[180] H. Shan and A. Banerjee. Generalized probabilistic matrix factorizations for collaborative filtering. In *Proceedings of the 10th IEEE International Conference on Data Mining*, ICDM '10, pages 1025–1030, 2010. DOI: 10.1109/ICDM.2010.116. 50, 80

[181] H. Shan and A. Banerjee. Residual bayesian co-clustering for matrix approximation. In *Proceedings of the 10th SIAM Conference on Data Mining*, SDM '10, pages 223–234, 2010. DOI: 10.1137/1.9781611972801.20. 71

[182] Shang Shang, Pan Hui, Sanjeev R. Kulkarni, and Paul W. Cuff. Wisdom of the crowd: Incorporating social influence in recommendation models. In *Proceedings of the 17th IEEE International Conference on Parallel and Distributed Systems*, ICPADS '11, pages 835–840, 2011. DOI: 10.1109/ICPADS.2011.150. 143

[183] Amit Sharma and Dan Cosley. Do social explanations work?: studying and modeling the effects of social explanations in recommender systems. In *Proceedings of the 22nd International Conference on World Wide Web*, WWW '13, pages 1133–1144, 2013. 143

[184] Rashmi R. Sinha and Kirsten Swearingen. Comparing recommendations made by online systems and friends. In *DELOS Workshop: Personalisation and Recommender Systems in Digital Libraries*, 2001. 127

[185] N. Srebro and T. Jaakkola. Weighted low rank approximation. In *Proceedings of the 20th International Conference on Machine Learning*, ICML '03, 2003. 19

[186] N. Srebro, J.D. M. Rennie, and T.S. Jaakola. Maximum-margin matrix factorization. In *Proceedings of the annual conference on Advances in Neural Information Processing Systems*, NIPS '05, pages 1329–1336, 2005. 20

[187] D.H. Stern, R. Herbrich, and T. Graepel. Matchbox: large scale online bayesian recommendations. In *Proceedings of the 18th International Conference on World Wide Web*, WWW '09, pages 111–120, 2009. DOI: 10.1145/1526709.1526725. 112

[188] M. Steyvers and T. Griffiths. *Latent Semantic Analysis: A Road to Meaning*, chapter Probabilistic topic models. Laurence Erlbaum, Mahwah, New Jersey, USA, 2007. 45, 62

[189] A. Strehl and J. Ghosh. Value-based customer grouping from large retail data-sets. In *Proc. of the SPIE Conference on Data Mining and Knowledge Discovery: Theory, Tools, and Technology*, pages 33–42, 2000. 12, 13

[190] Lei Tang and Huan Liu. *Community Detection and Mining in Social Media*. Synthesis Lectures on Data Mining and Knowledge Discovery. Morgan-Claypool, 2010. 138

[191] Y. W. Teh, D. Newman, and M. Welling. A collapsed variational bayesian inference algorithm for latent dirichlet allocation. In *Proceedings of the annual conference on Advances in Neural Information Processing Systems*, volume 6 of *NIPS '06*, pages 1378–1385, 2006. 85

[192] N.C Tewari, H.M. Koduvely, S. Guha, A. Yadav, and G. David. Mapreduce implementation of variational bayesian probabilistic matrix factorization algorithm. In *Proceedings of the 2013 IEEE International Conference on Big Data*, pages 145–152, 2013. DOI: 10.1109/BigData.2013.6691747. 84

[193] S. Vargas and P. Castells. Rank and relevance in novelty and diversity metrics for recommender systems. In *Proceedings of the 5th ACM Conference on Recommender Systems*, RecSys' 11, pages 109–116, 2011. DOI: 10.1145/2043932.2043955. 149

[194] S. Vargas and P. Castells. Exploiting the diversity of user preferences for recommendation. In *Proceedings of the 10th Conference on Open Research Areas in Information Retrieval*, OAIR '13, pages 129–136, 2013. 104, 109, 149

[195] Hanna M. Wallach. Topic modeling: Beyond bag-of-words. In *Proceedings of the 23rd International Conference on Machine Learning*, ICML '06, pages 977–984, 2006. DOI: 10.1145/1143844.1143967. 118, 120

[196] P. Wang, C. Domeniconi, and K.B. Laskey. Latent dirichlet bayesian co-clustering. In *Proceedings of the European Conference on Machine learning and Knowledge Discovery in Databases*, ECML/PKDD '09, pages 522–537, 2009. DOI: 10.1007/978-3-642-04174-7_34. 70

[197] M Weimer, A. Karatzoglou, and A.J. Smola. Improving maximum margin matrix factorization. In *Proceedings of the European Conference on Machine learning and Knowledge Discovery in Databases*, volume 5211 of *Lecture Notes in Computer Science*, 2008. DOI: 10.1007/978-3-540-87479-9_12. 20, 56

[198] C. Williams, B. Mobasher, and R. Burke. Defending recommender systems: detection of profile injection attacks. *Service Oriented Computing and Applications*, 1(3):157–170, 2007. DOI: 10.1007/s11761-007-0013-0. 10

[199] A. McCallum X. Wang and X. Wei. Topical n-grams: Phrase and topic discovery, with an application to information retrieval. In *Proceedings of the 7th IEEE International Conference on Data Mining*, ICDM '07, pages 697–702, 2007. DOI: 10.1109/ICDM.2007.86. 118

[200] L. Xiong, X. Chen, T. Huang, J.G. Schneider, and J.G. Carbonell. Temporal collaborative filtering with bayesian probabilistic tensor factorization. In *Proceedings of the 10th SIAM Conference on Data Mining*, SDM '10, pages 211–222, 2010. 125

[201] M. Xu, J. Zhu, and B. Zhang. Nonparametric max-margin matrix factorization for collaborative prediction. In *Proceedings of the annual conference on Advances in Neural Information Processing Systems*, NIPS '12, pages 64–72, 2012. 150

[202] L. Yao, D. Mimno, and A. McCallum. Efficient methods for topic model inference on streaming document collections. In *Proceedings of the 15th ACM SIGKDD International Conference on Knowledge Discovery and Data Mining*, KDD '09, pages 937–946, 2009. DOI: 10.1145/1557019.1557121. 150

[203] Mao Ye, Xingjie Liu, and Wang-Chien Lee. Exploring social influence for recommendation: a generative model approach. In *Proceedings of the 35th ACM SIGIR Conference on Research and Development in Information Retrieval*, SIGIR '12, pages 671–680, 2012. DOI: 10.1145/2348283.2348373. 143

[204] K. Yu, S. Zhu, J. Lafferty, and Y. Gong. Fast nonparametric matrix factorization for large-scale collaborative filtering. In *Proceedings of the 32nd ACM SIGIR Conference on Research and Development in Information Retrieval*, SIGIR '09, pages 211–218, 2009. DOI: 10.1145/1571941.1571979. 150

[205] P. Zezula, G. Amato, V. Dohnal, and M. Batko. *Similarity Search: The Metric Space Approach (Advances in Database Systems)*. Springer-Verlag, Secaucus, NJ, USA, 2005. 16

[206] K. Zhai, J. Boyd-Graber, N. Asadi, and M.L. Alkhouja. Mr. lda: a flexible large scale topic modeling package using variational inference in mapreduce. In *Proceedings of the 21st International Conference on World Wide Web*, WWW '12, pages 879–888, 2012. DOI: 10.1145/2187836.2187955. 84

[207] M. Zhang and N. Hurley. Avoiding monotony: improving the diversity of recommendation lists. In *Proceedings of the 2rd ACM conference on Recommender systems*, RecSys '08, pages 123–130, 2008. DOI: 10.1145/1454008.1454030. 101, 108

[208] M Zhang and N. Hurley. Novel item recommendation by user profile partitioning. In *Proceedings of the IEEE/WIC/ACM International Joint Conference on Web Intelligence and Intelligent Agent Technology*, WI-IAT '09, pages 508–515, 2009. DOI: 10.1109/WI-IAT.2009.85. 108

[209] T. Zhou, H. Shan, A. Banerjee, and G. Sapiro. Kernelized probabilistic matrix factorization: Exploiting graphs and side information. In *Proceedings of the 12th SIAM International Conference on Data Mining*, pages 403–414, 2012. 50

[210] Haiyi Zhu, Bernardo Huberman, and Yarun Luon. To switch or not to switch: understanding social influence in online choices. In *Proceedings of the SIGCHI Conference on Human Factors in Computing Systems*, CHI '12, pages 2257–2266, 2012. DOI: 10.1145/2207676.2208383. 129

[211] C. Ziegler, S. M. McNee, J.A. Konstan, and G. Lausen. Improving recommendation lists through topic diversification. In *Proceedings of the 14th International Conference on World Wide Web*, WWW '05, pages 22–32, 2005. DOI: 10.1145/1060745.1060754. 107, 108

[212] C.L. Zitnick and T. Kanade. Maximum entropy for collaborative filtering. In *Proceedings of the 20th Conference in Uncertainty in Artificial Intelligence*, UAI '04, pages 636–643, 2004. 28

Authors' Biographies

NICOLA BARBIERI

Nicola Barbieri is a post-doc in the *Web Mining* research group at Yahoo Labs – Barcelona. He graduated with full marks and honor and received his Ph.D. in 2012 at University of Calabria–Italy. Before joining Yahoo in 2012, he was a fellow researcher at ICAR-CNR. His research focuses on the development of novel data mining and machine learning techniques with a wide range of applications in social influence analysis, viral marketing, and community detection.

GIUSEPPE MANCO

Giuseppe Manco received a Ph.D. degree in computer science from the University of Pisa. He is currently a senior researcher at the Institute of High Performance Computing and Networks (ICAR-CNR) of the National Research Council of Italy and a contract professor at University of Calabria, Italy. He has been contract researcher at the CNUCE Institute in Pisa, Italy. His current research interests include knowledge discovery and data mining, Recommender systems, and Social Network analysis.

ETTORE RITACCO

Ettore Ritacco is a researcher at the Institute of High Performance Computing and Networks (ICAR-CNR) of the National Research Council of Italy. He graduated summa cum laude in Computer Science and received his Ph.D. in the doctoral school in System Engineering and Computer Science (cycle XXIII), 2011, at University of Calabria, Italy. His research focuses on mathematical tools for knowledge discovery, business intelligence and data mining. His current interests are Recommender Systems, Social Network analysis, and mining complex data in hostile environments.

Printed in the United States
by Baker & Taylor Publisher Services